Lecture Notes in Artificial Intelligence 11763

Subseries of Lecture Notes in Computer Science

Series Editors

Randy Goebel
  University of Alberta, Edmonton, Canada
Yuzuru Tanaka
  Hokkaido University, Sapporo, Japan
Wolfgang Wahlster
  DFKI and Saarland University, Saarbrücken, Germany

Founding Editor

Jörg Siekmann
  DFKI and Saarland University, Saarbrücken, Germany

More information about this series at http://www.springer.com/series/1244

Davide Calvaresi · Amro Najjar ·
Michael Schumacher · Kary Främling (Eds.)

Explainable, Transparent Autonomous Agents and Multi-Agent Systems

First International Workshop, EXTRAAMAS 2019
Montreal, QC, Canada, May 13–14, 2019
Revised Selected Papers

 Springer

Editors
Davide Calvaresi (iD)
University of Applied Sciences
of Western Switzerland (HES-SO)
Sierre, Switzerland

Michael Schumacher (iD)
University of Applied Sciences
of Western Switzerland (HES-SO)
Sierre, Switzerland

Amro Najjar (iD)
University of Luxembourg
Esch-sur-Alzette, Luxembourg

Kary Främling (iD)
Umeå University
Umeå, Sweden

ISSN 0302-9743   ISSN 1611-3349   (electronic)
Lecture Notes in Artificial Intelligence
ISBN 978-3-030-30390-7   ISBN 978-3-030-30391-4   (eBook)
https://doi.org/10.1007/978-3-030-30391-4

LNCS Sublibrary: SL7 – Artificial Intelligence

This Springer imprint is published by the registered company Springer Nature Switzerland AG
The registered company address is: Gewerbestrasse 11, 6330 Cham, Switzerland

Preface

In the last decade, the dependency of our society towards intelligent systems has dramatically escalated. For example, data-driven techniques (typical of machine learning - ML) such as classification, regression, and clustering are increasingly employed in semi-automated diagnosis based on medical image recognition, financial market forecasting, and customer profiling.

The internal mechanisms for such systems are opaque and not understandable nor explainable. Therefore, humans are still intertwined with those semi-automated intelligent systems due to legal, ethical, and user-requirement reasons. Moreover, the real world is characterized by uncountable and heterogeneous (possibly abstract) stimuli concurring in the composition of complex/articulated information. Persisting with specific and isolated solutions (heavily demanding continuous handcrafting) would only enhance the already unsustainable human-overhead.

To cope with the real world heterogeneity and enforce AI/ML trustworthiness, the recently born discipline named eXplainable Artificial Intelligence (XAI) moved towards Distributed Artificial Intelligence (DAI) approaches. In particular, the research community is fostering the adoption of Multi-Agent Systems (MAS) embodying DAI (e.g., autonomous vehicles, robots, smart buildings, and IoT) to enforce explainability, transparency, and (above all) trustworthiness.

This volume contains a selection of the extended papers presented at the First International Workshop on EXplainable TRansparent Autonomous Agents and Multi-Agent Systems (EXTRAAMAS 2019), held in conjunction with the International Conference on Autonomous Agents and Multi-Agent Systems (AAMAS 2019), during May 13–14, 2019, in Montreal, Canada.

The EXTRAAMAS 2019 organizers would like to thank the publicity chairs and Program Committee for their valuable work.

June 2019

Davide Calvaresi
Amro Najjar
Michael Schumacher
Kary Främling

Organization

General Chair

Davide Calvaresi   University of Applied Sciences Western Switzerland,
                    Switzerland
Amro Najjar   University of Luxembourg, Luxembourg
Michael Schumacher University of Applied Sciences Western Switzerland,
                    Switzerland
Kary Främling   Umeå University, Sweden

Publicity Chairs

Yazan Mualla   University Bourgogne Franche-Comté, France
Timotheus Kampik   Umeå University, Sweden
Amber Zelvelder   Umeå University, Sweden

Program Committee

Andrea Omicini   Università di Bologna, Italy
Ofra Amir   Technion IE&M, Israel
Joost Broekens   TU Delft, The Netherlands
Olivier Boissier   ENS, Mines Saint-Étienne, France
J. Carlos N. Sanchez   Umeå University, Sweden
Tathagata Chakraborti   IBM Research AI, USA
Salima Hassas   Lyon 1, France
Gauthier Picard   EMSE Saint-Étienne, France
Jean-Guy Mailly   Laboratoire d'Informatique de Paris Descartes, France
Aldo F. Dragoni   Università Politecnica delle Marche, Italy
Patrick Reignier   LIG Grenoble, France
Stephane Galland   UTBM, France
Laurent Vercouter   INSA Rouen, France
Helena Lindgren   Umeå University, Sweden
Grégory Bonnet   University of Caen, France
Jean-Paul Calbimonte   University of Applied Sciences Western Switzerland,
                        Switzerland
Sarath Sreedharan   Arizona State University, USA
Brent Harrison   Georgia Institute of technology, USA
Koen Hindriks   VU, The Netherlands
Laëtitia Matignon   University Lyon 1, France
Simone Stumpf   London City University, UK
Michael W. Floyd   Knexus Research, USA
Kim Baraka   CMU, USA

Contents

Explanation and Transparency

Towards a Transparent Deep Ensemble Method Based
on Multiagent Argumentation 3
    Naziha Sendi, Nadia Abchiche-Mimouni, and Farida Zehraoui

Effects of Agents' Transparency on Teamwork..................... 22
    Silvia Tulli, Filipa Correia, Samuel Mascarenhas, Samuel Gomes,
    Francisco S. Melo, and Ana Paiva

Explainable Robots

Explainable Multi-Agent Systems Through Blockchain Technology........ 41
    Davide Calvaresi, Yazan Mualla, Amro Najjar, Stéphane Galland,
    and Michael Schumacher

Explaining Sympathetic Actions of Rational Agents 59
    Timotheus Kampik, Juan Carlos Nieves, and Helena Lindgren

Conversational Interfaces for Explainable AI:
A Human-Centred Approach 77
    Sophie F. Jentzsch, Sviatlana Höhn, and Nico Hochgeschwender

Opening the Black Box

Explanations of Black-Box Model Predictions by Contextual
Importance and Utility...................................... 95
    Sule Anjomshoae, Kary Främling, and Amro Najjar

Explainable Artificial Intelligence Based Heat Recycler Fault Detection
in Air Handling Unit....................................... 110
    Manik Madhikermi, Avleen Kaur Malhi, and Kary Främling

Explainable Agent Simulations

Explaining Aggregate Behaviour in Cognitive Agent Simulations
Using Explanation... 129
    Tobias Ahlbrecht and Michael Winikoff

BEN: An Agent Architecture for Explainable and Expressive
Behavior in Social Simulation................................ 147
    Mathieu Bourgais, Patrick Taillandier, and Laurent Vercouter

Planning and Argumentation

Temporal Multiagent Plan Execution: Explaining What Happened. 167
  Gianluca Torta, Roberto Micalizio, and Samuele Sormano

Explainable Argumentation for Wellness Consultation 186
  Isabel Sassoon, Nadin Kökciyan, Elizabeth Sklar, and Simon Parsons

Explainable AI and Cognitive Science

A Historical Perspective on Cognitive Science and Its Influence
on XAI Research 205
  Marcus Westberg, Amber Zelvelder, and Amro Najjar

Author Index 221

Explanation and Transparency

Towards a Transparent Deep Ensemble Method Based on Multiagent Argumentation

Naziha Sendi[1,2](✉) [ID], Nadia Abchiche-Mimouni[1] [ID], and Farida Zehraoui[1] [ID]

[1] IBISC, Univ Evry, Université Paris-Saclay, 91025 Evry, France
nsendi@visiomed-lab.fr
[2] Bewell Connect, 75016 Paris, France

Abstract. Ensemble methods improve the machine learning results by combining different models. However, one of the major drawbacks of these approaches is their opacity, as they do not provide results explanation and they do not allow prior knowledge integration. As the use of machine learning increases in critical areas, the explanation of classification results and the ability to introduce domain knowledge inside the learned model have become a necessity. In this paper, we present a new deep ensemble method based on argumentation that combines machine learning algorithms with a multiagent system in order to explain the results of classification and to allow injecting prior knowledge. The idea is to extract arguments from classifiers and combine the classifiers using argumentation. This allows to exploit the internal knowledge of each classifier, to provide an explanation for the decisions and facilitate integration of domain knowledge. The results demonstrate that our method effectively improves deep learning performance in addition to providing explanations and transparency of the predictions.

Keywords: Deep learning · Ensemble methods · Knowledge extraction · Multiagent argumentation

1 Introduction

The Machine Learning (ML) models have started penetrating into critical areas like health care, justice systems, and financial industry [30]. Thus, explaining how the models make the decisions and make sure the decision process is aligned with the ethical requirements or legal regulations becomes a necessity. Ensemble learning methods as a ML model have been widely used to improve classification performance in ML. They are designed to increase the accuracy [42] of a single classifier by training several different classifiers and combining their decisions to output a single class label, such as Bagging [10] and Boosting [18]. However, traditional ensemble learning methods are considered as black boxes. In fact, they mainly use weighted voting to integrate multiple base classifiers. Although

© Springer Nature Switzerland AG 2019
D. Calvaresi et al. (Eds.): EXTRAAMAS 2019, LNAI 11763, pp. 3–21, 2019.
https://doi.org/10.1007/978-3-030-30391-4_1

this "majority-voting" approach is relatively simple, it lacks interpretability for users. In addition, only classification results of base classifiers are integrated, rather than their internal classification knowledge.

In order to overcome the weaknesses of traditional ensemble learning methods, we introduce a new ensemble method based on multiagent argumentation. The integration of argumentation and ML has been proven to be fruitful [11] and the use of argumentation is an intelligent way of combining learning algorithms since it can imitate human decision-making process to realize the conflict resolution and knowledge integration and also provide explanation behind decisions.

Due to the above advantages of argumentation, this paper proposes a transparent ensemble learning approach, which integrates multiagent argumentation into ensemble learning. The idea is to construct for each instance arguments for/against each classifier, evaluate them, and determine among the conflicting arguments the acceptable ones. The arguments will be extracted automatically from classifiers. As a result, a valid classification of the instance is chosen. Thus, not only the class of the instance is given, but also the reasons behind that classification are provided to the user as well in a form that is easy to grasp.

The article is organized as follows. Section 2 introduces the related research works about ensemble learning, argumentation and explainable intelligent systems. Section 3 presents the principles of our method. A realization process of a specific use case is given in Sect. 4. Experimental results on public datasets are presented and discussed in Sect. 5. Finally, Sect. 6 summaries the contributions and future works.

2   State of Art

Recent years, ensemble learning, argumentation technology and explainable intelligent systems attract much attention in the field of Artificial Intelligence (AI). In this section, we focus on the related works about ensemble learning, argumentation in ML and Explainable artificial systems.

2.1   Ensemble Method

Many ways for combining base learners into ensembles have been developed. Dieterich [14] classifies ensemble methods based on the way the ensembles are built from the original training dataset. We describe the four most used strategies. The first method manipulates the training examples to generate multiple hypotheses. The learning algorithm is executed several times, each time with a different subset of the training examples. These methods include the most popular ensemble methods: bagging [10] and boosting (AdaBoost) [18]. Bagging builds learning algorithms based on bootstrap samples. Boosting associates, for each classifier, a set of weights to the training examples based on previous classification results. The second technique for constructing an ensemble of classifiers is to manipulate the outputs values (classes) that are given to the learning algorithm. One of the most used approaches is called error correcting output coding

[15]. It starts by randomly partitioning the classes into two subsets. Then a new learning problem can be constructed using the two new classes (each class represents a subset of the initial classes). The input data are then relabelled and this process is repeated recursively generating different subsets in order to obtain an optimal ensemble of classifiers. The third type of approaches manipulates the input features available in the training set. The input features are divided into several parts. Each part corresponds to a coherent group of features. This allows to obtain diverse classifiers that use different features types. The last method is to apply randomized procedures to the learning processes in order to improve the diversity. For example, in the backpropagation algorithm for training neural networks the weights of the network are initialized randomly. The algorithm is applied to the same training examples but with different initial weights. The resulting classifiers and their results can be quite different. We can also distinguish ensemble fusion methods from ensemble selection methods based on the way the outputs of the ensemble algorithms are combined. Ensemble fusion methods [36] like Fuzzy fusion and majority voting, combine all the outputs of the base classifiers by the majority voting and the class that collects most votes is predicted by the ensemble while ensemble selection methods [36] like test and select methods and cascading classifiers choose the best base classifier among the set of base learners for a specified input, and the output of the ensemble is the output of the selected best classifier. Most of research works has focused on the advantages of ensemble methods to improve the algorithm's performance. However, one of their major drawback is their lake of transparency, since no explanation of their decisions has been offered.

2.2   Argumentation in ML

Several approaches have been proposed to combine multiagent argumentation and ML. Hao et al. [23] present Arguing Prism, an argumentation based approach for collaborative classification which integrates the ideas from modular classification inductive rules learning and multiagent dialogue. Each participant agent has its own local repository (data instances) and produces reasons for or against certain classifications by inducing rules from their own datasets. The agents use argumentation to let classifiers, learned from distributed data repositories, reaching a consensus rather than voting mechanisms. This approach is interesting because it allows avoiding simple voting and generates arguments in a dynamic way. Unfortunately its use is restricted to decision trees. Wardeh et al. [43] present a classification approach using a multiagent system founded on an argumentation from experience. The technique is based on the idea that classification can be conducted as a process whereby a group of agents argue about the classification of a given case according to their experience which is recorded in individual local data sets. The arguments are constructed dynamically using classification association rule mining [2] techniques. Even if this approach argues for the use of local data for the argument exchange between the agents, there is a chairperson agent which acts as a mediator agent for the coordination of the whole multiagent system. From our point of view, this is a weak point, since the system fails to perform any classification if the chairperson agent fails. A more recent work [40] presents preliminary experiments confirming the necessity of the combination of ML and argumentation. The authors propose to bring the gap between ML and knowledge representation and reasoning and suggest that this can be applied to multiagent argumentation.

Existing approaches differ in their use of argumentation and in their choice of argumentation framework/method. Finally, different approaches achieve different and desirable outcomes, ranging from improving performances to rendering the ML process more transparent by improving its explanatory power. These works illustrate the importance of building arguments for explaining ML examples. But all of them are dedicated to rule association and are used in monolotihic way. The most important point is that, none of them addresses deep learning methods, despite these are among the most powerful ML algorithms. The use of argumentation techniques allows to obtain classifiers, which can explain their decisions, and therefore addresses the recent need for explainable AI: classifications are accompanied by a dialectical analysis showing why arguments for the conclusion are preferred to counterarguments.

2.3   Explainable Intelligent Systems

Explainable artificial intelligence (XAI) has been gaining increasing attention. XAI aims to make AI systems results more understandable to humans. Most of the existing works in literature focus on explainability in ML which is just one field of AI. However, the same issues also confront other intelligent systems. Particularly explainable agent are beginning to gain recognition as a promising derived field of XAI [1].

Explainability in ML. In the studied literature, many interpretable models are used such as linear models, decision trees and logic rules [21] to explain blackboxes. In order to provide a flexible and structured knowledge with textual representation, we have used logic rules. The advantage of this representation is that it facilitates the integration/injection of prior knowledge.

In recent years, many approaches for rule extraction from trained neural networks have been developed. According to Andrews et al. [4], the techniques of the rule extraction can be grouped into three main approaches namely decompositional, pedagogical and eclectic. The decompositional approach [19,38,44] extracts the symbolic rules by analyzing the activation and weights of the hidden layers of the neural network. The pedagogical approach [5,13] extracts rules that represent input-output relationship so as to reproduce how the neural networks learned the relationship. The eclectic approach [7,25,31] is a combination of the decompositional and the pedagogical approaches.

Tran and Garcez [41] propose the first rule extraction algorithm from Deep Belief Networks. However, these stochastic networks behave very differently from the multilayer perceptrons (MLP), which are deterministic. Zilke et al. [44] have proposed an algorithm that uses a decompositional approach for extracting rules from deep neural networks called DeepRED. This algorithm is an extension of the CRED algorithm [32]. For each class, It extracts rules by going through the hidden layers in descending order. Then, it merges all the rules of a class in order to obtain the set of rules that describes the output layer based on the inputs. Bologna et al. [7] proposed a Disretized Interpretable Multilayer Perceptron (DIMLP) that uses an eclectic approach to represent MLP architectures. It estimates discriminant hyperplanes using decision trees. The rules are defined by the paths between the root and the leafs of the resulting decision trees. A pruning strategy was proposed in order to reduce the sets of rules and premises.

Craven et al. [12] propose a method to explain the behavior of a neural network by transforming rule extraction into a learning problem. In other words, it consists in testing if an input from the original training data with its outcome is not covered by the set of rules, then a conjunctive (or m-of-n) rule is formed from considering all the possible antecedents. The procedure ends when all the target classes have been processed.

Explainability in MAS. As agents are supposed to represent human behavior, works in this area mainly focus on behavior explanations generation so that agents could explain the reasons behind their actions. Harbers et al. [24] propose to use self-explaining agents, which are able to generate and explain their own behavior for the training of complex, dynamic tasks in virtual training. To give trainees the opportunity to train autonomously, intelligent agents are used to generate the behavior of the virtual players in the training scenario. The explanations aim to give a trainee an insight into the perspectives of other players, such as their perception of the world and the motivations for their actions, and thus facilitate learning. Johnson et al. propose a system called Debrief [26], which has been implemented as part of a fighter pilot simulation and allows trainees to ask an explanation about any of the artificial fighter pilot's actions. To generate an answer, Debrief modifies the recalled situation repeatedly and systematically, and observes the effects on the agent's decisions. With the observations, Debrief determines what factors were responsible for 'causing' the decisions. VanLent et al. [29] describes an AI architecture and associated explanation capability used by Full Spectrum Command, a training system developed for the U.S. Army by commercial game developers and academic researchers. The XAI system has been incorporated into a simulation based training for commanding a light infantry company. After a training session, trainees can select a time and an entity, and ask questions about the entity's state. However, the questions involve the entity's physical state, e.g. its location or health, but not its mental state. Javier et al. [20] describe a natural language chat interface that enables the vehicle's behaviour to be queried by the user. The idea is to obtain an interpretable model of autonomy by having an expert "speak out-loud" and provide explanations during a mission. To provide simplistic explanations for the Nonplayer characters in video games, Molineaux et al. [35] consider a new design for agents that can learn about their environments, accomplish a range of goals, and explain what they are doing to a supervisor.

3   Method

We propose an original method, based on multiagent argumentation, which combines several DNNs. This way of combining DNNS allows not only to provide explanations of individual predictions, but also the injection of domain knowledge. So, the argumentation process uses knowledge extracted from the individual classifiers and domain knowledge. The method proceeds in two main phases: (1) arguments extraction phase and (2) multiagent argumentation phase (see Fig. 1).

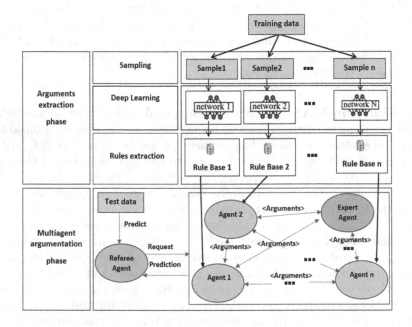

Fig. 1. Approach architecture

3.1   Arguments Extraction Phase

The classifiers are built using bootstrap training samples which are generated from the training dataset. As in bagging ensemble method [10], a bootstrap sample is obtained by a random selection of examples with replacement from the original training dataset.

Deep multilayer network (DMLP) is used as base classifier for the ensemble method. A DMLP consists of an input layer that receives input examples, hidden layers that are fully connected to the previous and the next layers and an output layer that provides the network outputs. These consist of the probabilities of an input example to belong to the classes. Let's h_i^l the ith neuron of the hidden layer l, its activation is defined by: $h_i^l = f(\sum_j w_{ji}^l h_j^{l-1})$, where w_{ji}^l is the weight of the connection from the jth neuron of the layer $(l-1)$ to the ith neuron of the layer l (h^0 represents the input layer) and f is the activation function. For the hidden layers, we used the Rectified Linear Units ($ReLU$) activation function, which gives good results in practice. It is defined as follows: $ReLU(x) = max(0, x)$. We used the softmax activation function for the output layer in order to obtain the probabilities that the input $X = (x_1, x_2, ..., x_n)$ belongs to a class c. This function is defined by: $softmax(h_c^o) = \frac{exp^{h_c^o}}{\sum_l exp^{h_l^o}}$.

To train the DMLP, we used the adam [27] optimizer and the cross-entropy cost function L, which is the best choice in state-of-the art implementations. It is defined by: $L(Y, O) = -\frac{1}{N} \sum_i \sum_l y_{il} \ln(o_{il})$.

Rules extraction step is very important since it allows to explain the predictions and to make the link between the classifiers and the MAS. To extract classification rules from DNNs, we have evaluated one pedagogical approach [13] and one eclectic approach [7]. We have chosen these approaches because they are scalable and adapted to the use of multiple deep learning algorithms. In [13], the authors proposed an algorithm (TREPAN) that extracts rules from DNN by producing a decision tree. TREPAN uses a best-first procedure when building the decision tree. This consists in choosing the node that most increases the fidelity of the tree. A rule is defined by the path between the root and a leaf of the built tree. In [7], the first phase consists in approximating the discriminant frontier built by a DMPL using discriminant hyperplanes frontiers. In the second phase, a Discretized Interpretable Multilayer Perceptron (DIMLP) model is extracted based on the discriminant hyperplanes frontiers. The multilayer architecture of DIMLP is more constrained than that of a standard multilayer perceptron. From this particular constrained network, rules based on continuous attributes can be extracted in polynomial time. The extracted classification rules from each classifier constitute a rule base that is associated to the classifier. Each rule base is then embedded in an agent.

The form of a classification rule CR is: $CR : (pr_1) (pr_i) ... (pr_n) \implies (class(CR) = c, confidence\_score(CR) = s)$, where: $pr_i \in premises(CR)$ $(1 \leq i \leq n)$ are the premises of the rule CR that the example must satisfy to be classified in $c \in C$ (C is the set of classes). The form of the premise pr_i is defined by $pr_i = (x_i \ op \ \alpha_i)$ where x_i is the value of the ith attribute, α_i is a real number and op is an operator. s $(0 \leq s \leq 1)$ is a confidence score that is associated to the rule CR. This score depends on the number $ne_c^+(CR)$ of examples that are well classified by the rule CR. To take into account the fact that most real datasets are unbalanced the number of well classified examples $ne_c^+(CR)$ is divided by the total number of examples ne_c in the class c : $confidence\_score(CR) = \frac{ne_c^+(CR)}{ne_c}$.

Domain knowledge is also modeled in the form of rules, named expert rules (ERs):

$ER : (pr_1) \ (pr_i) \ ... \ (pr_n) \implies (class(ER) = c)$, where $pr_i \in premises(ER)$ $(1 \leq i \leq n)$ are the premises of the rule ER that the example must satisfy to be classified in the class $c \in C$ based on the official experts' knowledge. For example, the ER_1 rule below expresses that an official recommendation for hypertension is to prefer the beta blockers (BB) treatment for young people: $ER_1 : (age < 50) \implies (class(ER_1) = BB)$.

As said earlier, each rule base is encapsulated in an agent. In order to allow injecting prior knowledge in the system, an *Expert* agent is added for embedding the knowledge base which models prior knowledge provided by domain experts.

3.2   Multiagent Argumentation Phase

Abstract argumentation has been introduced by Dung [16] for modeling arguments by a directed graph, where each node represents an argument and each arc denotes an attack by an argument on another. To express that an argument a attacks an argument b (that is, argument a is stronger than b and b is discarded), a binary relation is defined. The graph is analyzed to determine which set of arguments is acceptable according to general criteria. Structured argumentation has been introduced by [6] to formalize arguments in such a way that premises and claim (such as a class for a CR, see Sect. 3.1) of the argument are made explicit, and the relationship between premises and claim is formally expressed (for instance using rule deduction). In our case, the arguments need to be structured since they are already given in the form of rules. Several works propose semantics for determining acceptability of arguments. Preferences based approaches consider global evaluation of the arguments (extensions). Since in our distributed approach it is hard to use a global preference based argumentation, we exploited local (agent) preference based method. As we will see later, the score and the number of premises of the rules are used during the encounter of arguments. [28] distinguishes dialogical argumentation where several agents exchange arguments and counterarguments in order to argue for their opinion, from monological argumentation whose emphasis is on the analysis of a set of arguments and counterarguments.

Modelling the argumentation process consists in allowing each agent of the MAS to argue for its own prediction against other agents. So, we have focused on dialogical argumentation for the implementation of the argumentation process [37]. More precisely, agents engage in a process of persuasion dialogue [22] since they have to convince other agents that their prediction is better. Through the argumentation process, each agent uses the rules of its embedded rule base to answer to a prediction request and to provide arguments during the argumentation process. Since all the agents are able to participate to the argumentation process by exchanging messages, we have focused on multilateral argumentative dialogues protocols [9]. According to [34], multilateral argumentative dialogue protocol (MADP) is based on several rules that are instanciated in our approach as explained hereafter. Moreover, it has been shown in [3] that agents role affect positively the argumentation process. So, in order to organize the dialogue, four distinct agent roles are defined:

(i) Referee agent: broadcasts the prediction request and manages the argumentation process;
(ii) Master: agent that answers first to the *Referee* request;
(iii) Challenger: agent who challenges the *Master* by providing arguments;
(iv) Spectator: agent who does not participate to the argumentation process.

The *Referee* is an "artifact" agent role that is assigned in a static way. This agent interacts with the user for acquiring the prediction request and collecting the final result. The argumentation process is performed through agents communication. For that purpose, we adopted speech acts language [39]. Let X be the input data, where X is a vector of attributes values $(x_i)_{i=1,...,n}$, c the class to predict. Three kinds of agents are present in the MAS: A_r the Referee Agent, A_e the Expert Agent who embeeds the ERs, the agents which embeeds the CRs (A_m is the agent whose role is Master and A_c the agent whose role is Challenger). Seven communication performatives are used to instanciate the rules of the MADP as follows:

1. **Starting rules**: the dialogue starts as soon as the user asks for a prediction. A_r uses the REQUEST performative to brodcast the request for a prediction. The content of the message is: $(X, ?c)$.
2. **Locution rules**: an agent A_i sends an information by using the INFORM performative and asks for an information by using the ASK performative.
3. **Commitment rules**: two rules are defined. The first one manages the prediction request by using the PROPOSE performative, allowing an agent A_i to propose an opinion by selecting the best rule that matches the request: $R_x^{i*} \in RB_x^i$ such that $confidence\_score(R_x^{i*}) = \max_{R_x^i \in RB_x^i} (confidence\_score(R_x^i))$, where $RB_x^i = \{R^i : R^i \in RB^i \wedge premises(R^i) \subset x\}$ (RB^i is the rule base associated to the agent A_i). The second rule allows an agent to declare its defeat by using the DEFEAT performative.
4. **Rules for combination of commitments**: three rules for dealing with COUNTER, DISTINGUISH, CHECK performatives are defined. They define how acceptance or rejection of a given argument is performed. A_c uses the COUNTER speech act to attack the argument of A_m (associated to the rule R_x^{m*}) by selecting the rule R_x^{c*} such that $confidence\_score(R_x^{c*}) > confidence\_score(R_x^{m*})$. A_c uses the DISTINGUISH speech act to attack the opponent's argument, in case of equality of rule scores of A_c and A_m, they use the number of premises in their proposed rules as arguments: If $premise\_number(R_x^{c*}) > premise\_number(R_x^{m*})$ then A_c becomes the new Master ($premise\_number$ is the number premises of a rule). The expert agent A_e uses the CHECK speech act to check if the proposed rule R_x^{i*} by an agent A_i does not violate the rules $R_x^e \in RB^e$.
5. **Termination rules**: the dialogue ends when no agent has a rule to trigger.

Agents dialogues specification and behavior allow the agents to implement the described argumentation protocol. An opinion is a prediction which is performed by an agent. It is represented in the form of a rule (CR) as described in Sect. 3.1. According to argument domain vocabulary a CR is a structured argument. It is modeled in the form of a production rule with its associated score. The later is used in the argumentation process to encounter an argument. When a prediction is requested, the agent uses the rule which matches the request and uses the adequate performative to proceed to the argumentation process. We used Jess (Java Expert System Shell) language which is a rule engine and a scripting environment written entirely in Java. Jess deals with first order logic and allows to organize the rules into packages according to their use. Jess uses an enhanced version of the Rete [17] algorithm to manage priority of the rules. The argumentation process begins as soon as the Referee Agent broadcasts a request for a prediction and manages the dialogue process. Each agent produces an opinion by selecting the best rule that matches the request. Once an agent sends his opinion, the Referee Agent sends his proposed opinion to the Expert Agent for verification. Expert Agent checks if the opinion matches with the recommendations, then he sends a message to the Referee Agent to express his acceptance if there is no conflict with the expert knowledge else he sends a rejection. The first agent who offers an accepted opinion become the Master. Other agents can challenge the Master by forming a challengers queue; the first participant in the queue is selected by the Referee agent to be a Challenger. All other agents except the Master and the Challenger agents adopt the Spectator role. For each discussed opinion, the agents can produce arguments from their individual knowledge base. When a Master is defeated by a Challenger, the Challenger becomes the new Master, and then can propose a new opinion. It should be noted that the defeated argument of the old Master can not be used again, the old Master can only produce a new argument to apply for Master once more. Otherwise, if a Challenger is defeated, the next participant in the Challengers queue is selected as the new Challenger, and the argumentation continues. If all challengers are defeated, the Master wins the argumentation and the Master's winning rule is considered as the prediction of the system. If there is no agent applying for the Master role, the argumentation is stopped. Since the number of arguments produced by the participants is finite and the defeated arguments can not be allowed to use repeatedly, the termination of the argumentation process is guaranteed. As we will see in the Case-study section, the output of the MAS contains not only the winning prediction and its explanation, but also the whole dialogue path which led to the result.

4   Case Study

Ensemble Learning algorithms have advanced many fields and produced usable models that can improve productivity and efficiency. However, since we do not really know how they work, their use, specifically in medical problems is problematic. We illustrate here how our approach can help both physicians and patients to be more informed about the reasons of the prediction provided by the system.

4.1   Data Description

We have used a specific dataset that is a realistic virtual population (RVP) [33] with the same age, sex and cardiovascular risk factors profile than the French population aged between 35 and 64 years old. It is based on official French demographic statistics and summarized data from representative observational studies. Moreover, a temporal list of visits is associated to each individual. For the current experiments, we have considered 40000 individuals monitored for hypertension during 10 visits per individual. Each visit contains: the systolic blood pressure (SBP), diastolic blood pressure (DBP), class hypertension treatment, number of treatment changes etc. For hypertension treatment, 5 major classes of drugs have been considered: calcium antagonist (AC), beta blockers (BB), ACE inhibitors (ICE), diuritics (DI) and sartans (SAR). The data of the RVP have been used to predict the treatment changing with our MAS, following the steps described in the precedent sections. In order to lunch the experimentations, the MAS is built by encapsulating each rule base in an agent.

4.2   Scenarios Illustration

Tow scenarios illustrate the argumentation process: the first without domain knowledge injection and in the second, we have injected few medical recommendations.

Scenario 1: without prior knowldge injection uses three DMLPs. The architecture of DMLPs was determined empirically. The retained architecture contains two hidden layers, it consists of: 22 input neurons, 22 neurons in the first hidden layer, 20 in the second hidden layer, 6 output neurons (five neurons representing the drug classes and one neuron representing the patients with no treatment). The DMLPs was trained for 1500 epochs. Each DMLP generates is built using one bootstrap sample. We extracted knowledge bases from the three DMLPs using the eclectic rule extraction approach proposed in [7]. Extracting rules from neural networks allows to give an overview of the logic of the network and to improve, in some cases, the capacity of the network to generalize the acquired knowledge. Rules are very general structures that offer a form easy to understand when finding the right class for an example. Table 1 shows the properties of the three rule bases extracted from the three DMLPs in terms of number of rules per base, examples per rule, premises and premises per rule. For example the rule base RB^1 contains 182 extracted rules, it uses in total 96 different premises, the average number of premises per rule is about 7.7 and the average number of examples per rule is about 652.7. Each rule base is then embedded in an agent. The rules are thus considered as individual knowledge of the agents. When a prediction is requested, instead of predicting the treatment class by majority voting like in classical ensemble methods, each agent uses a rule, which matches the request, to argue with other agents in the MAS in order to provide the best prediction for the current request. The process of argumentation executes as described in Sect. 3. This is illustrated in the following example.

Table 1. Rule bases properties.

Rule bases properties	RB^1	RB^2	RB^3
Number of rules	182	351	256
Number of premises	96	106	88
Number of premises per rule	7.7	9.1	9.7
Number of examples per rule	652.7	752.9	395.4

Let be p_1 a patient that is described by the following attributes: p_1: $[age = 64][sex = female][Visit_0 : SBP = 132.2, DBP = 79.5][Visit_1 : SBP = 125.3, DBP = 87.1][Visit_2 : SBP = 117.8, DBP = 89.1][Visit_3 : SBP = 103.4, DBP = 84.7]$. p_1 should be treated by the treatment AC and the objective of the system is to predict this optimal treatment following the argumentation process described in Sect. 2. The possible negotiation arguments are the weight of the rules and their premises number. To simplify the current scenario, we consider only the confidence scores of the rules. At the beginning of the scenario, the Referee Agent broadcasts the requested prediction, that is predicting the optimal treatment for the patient p_1. Then each agent produces its opinion and asks for the Master role. Agent A_1 becomes the first Master and offers its opinion as follows: "this case should be in the class DI depending on the rule: $R_{p_1}^{1*}$: $(age > 54)$ $(DBP_{Visit1} > 77.4)$ $(DBP_{Visit2} > 85.1)$ $\implies (class(R_{p_1}^{1*}) = DI, confidence\_score(R_{p_1}^{1*}) = 0.61)$". Agent A_2 challenges agent A_1 using DistinguishRule as follows: "$R_{p_1}^{1*}$ is unreasonable because of rule $R_{p_1}^{2*}$: $(age > 61)$ $(DBP_{Visit0} > 142.5)$ $(DBP_{Visit1} > 77.1)$ $(DBP_{Visit1} > 77.1)$ $(SBP_{Visit3} > 109.5) \implies (class(R_{p_1}^{2*}) = BB, confidence\_score(R_{p_1}^{2*}) = 0.72)$". The confidence score of the rule $R_{p_1}^{2*}$ is higher than the one of $R_{p_1}^{1*}$ Agent A_1 can not propose any rule to attack Agent A_2 and admits that he is defeated. Then Agent A_2 becomes the new Master and offers his own opinion. The argumentation process continues until none of the agent is able to propose an opinion nor challenging another agent opinion. At the end, the master gives his prediction of the hypertension medication in a form easy to understand. In this case, the final prediction is made by the agent A_3: $R_{p_1}^{3*}$: $(age > 50)$ $(DBP_{Visit0} < 81.3)$ $(DBP_{Visit1} > 86.8)$ $(SBP_{Visit2} < 120.3)$ $(SBP_{Visit3} > 112.1) \implies (class(R_{p_1}^{3*}) = AC, confidence\_score(R_{p_1}^{3*}) = 0.81)$.

Scenario 2: With Prior Knowledge Injection. Injecting knowledge domain is very crucial for a decision making system. Medicine is one of the critical areas which needs the injection of recommendations for healthcare to improve the system reliability. In order to illustrate that, our approach improves the treatment prediction when adding domain knowledge. We have injected few medical recommendations for hypertension treatment into the expert agent A_e. Examples of medical recommendations are given bellow: (age < 50 years) $\implies BB$; (age > 50 years) $\implies DI$. The major role of A_e is to check if there is conflict between the proposed opinion and the expert knowledge. In this scenario, we have used three agents, each one contains extracted rule base from each DMLP and an extra agent which contains the expert knowledge. The process of argumentation executes as described in Sect. 3. This is illustrated in the following example.

Let be p_2 a patient that is described by the following attributes: p_2: $[age = 56][sex = female][Visit_0 : SBP = 112.2, DBP = 79.6][Visit_1 : SBP = 125.4, DBP = 89.7][Visit_2 : SBP = 103.7, DBP = 88.7][Visit_3 : SBP = 132.4, DBP = 81.7]$. p_2 should be treated by the treatment BB and the objective of the system is to predict this optimal treatment as illustrated in Fig. 2.

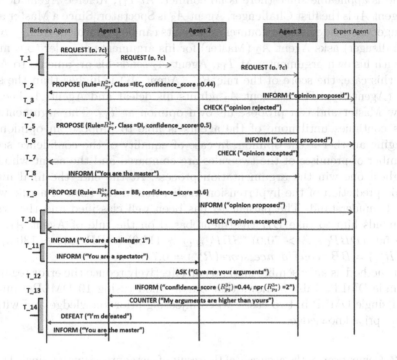

Fig. 2. Illustration of the case study argumentation process.

At the first iteration T_1, the Referee Agent broadcasts the prediction request by transmitting the attributes p_2 and the requested class $?c$ to predict. Each agent produces his opinion by selecting the best rule that matches the request. At T_2 Agent A_2 proposes his opinion as follows: "the requested class should be ICE based on the rule: $R_{p_2}^{2*}$: $(age > 50)$ $(SBP_{Visit1} > 101.4) \implies (class(R_{p_2}^{2*}) = ICE, confidence\_score(R_{p_2}^{2*}) = 0.44)$". At T_3, the Referee Agent sends the proposed opinion of Agent A_2 to the Expert Agent A_e for verification in order to check if the opinion matches with the recommendations. At T_4, Expert Agent A_e sends a message to the Referee Agent to express his rejection and declares that $confidence\_score(R_{p_2}^{2*})$ is inapplicable since the predicted class DI (given by this rule) does not match with the predicted class of the recommendation rule: R_1^e: $(age > 50 years) \implies (DI)$. At T_5, Agent A_3 proposes his opinion as follows: "the requested class should be BB based on the rule: R_{p2}^{3*}: $(age > 50)$ $(DBP_{Visit2} > 80.9)$ $(SBP_{Visit3} < 145) \implies (class(R_{p2}^{3*}) = BB,\ confidence\_score(R_{p2}^{3*}) = 0.56)$". At T_6, Referee Agent sends the suggested choice to the Expert agent for verification. A_e declares that this rule is applicable since there is no conflict at T_7. At T_8, Referee Agent declares that agent A_3 is defined as a Master.

At T_9, Agent A_1 proposes his opinion as follows:"the requested class should be BB based on the rule: R_{p2}^{1*}: $(age < 66)$ $(SBP_{Visit3} < 135.1)$ $(DBP_{Visit3} > 79)$ $\implies (class(R_{p2}^{1*}) = BB,\ confidence\_score(R_{p2}^{1*}) = 0.6)$". At T_{10}, A_e declares that this rule is applicable since there is no conflict. At T_{11}, Referee Agent declares that Agent A_1 is the first Challenger, Agent A_3 is Spectator. Since a Master and a Challenger are defined, the encounter arguments can be performed. At T_{12}, Agent A_1 (Challenger) asks Agent A_2 (Master) for his arguments in order to compare them with his own arguments. At T_{13}, Agent A_2 sends his arguments to Agent A_1. In this case, the score of the rule R_{p2}^{1*} (Agent A_1) is higher than the score of R_{p2}^{2*} (Agent A_2). Thus, Agent A_2 admits his defeat and Agent A_1 becomes the new Master and can propose his own opinion at T_{14}. The argumentation process continues until none of the agents is able to propose an opinion nor challenging another agent opinion. In case of equality of the confidence scores, the number of premises of the two rules are compared and the agent who have the highest one win the argumentation process. At the end, the final master gives his prediction of the hypertension medication in the form of a rule which is easy to understand. The patient p_2 has been well classified and the system recommends him to take BB treatment based on the rule of Agent A_1: R_{p2}^{1*}: $(age > 50)$ $(DBP_{Visit2} > 70.0)$ $(SBP_{Visit2} > 112.0)$ $(SBP_{Visit3} > 130.5) \implies (class(R_{p2}^{1*}) = BB,\ confidence\_score(R_{p2}^{1*}) = 0.72)$.

Our method as an ensemble method can effectively reduce the error regarding to a single DMLP. Table 2 shows that our method (using 10 DMLPs) outperforms a single DMLP in two cases: when injecting prior knowledge and without injecting prior knowledge.

Table 2. Comparison of the accuracy of the result of our approach with a single DMLP.

Single DMLP	Without prior knowledge injection	With prior knowledge injection
79.8 ± 0.01	83.2 ± 0.03	89.0 ± 0.01

As we can see in the Table 2, expert knowledge injection improves accuracy of classification. We improved the results classification and explained decision by providing not only comprehensible classification rule behind the decision but also the sent and received messages by the agents. So one can obtain a trace allowing to distinguish the unfolding communication between agents. In Fig. 2, bold arrows messages that lead to a final prediction treatment for p_2 patient. Moreover, our approach is able to exploit domain knowledge that control the system and gives trust to the expert.

5   Experimentation

In the experiments we used 10 datasets representing classification problems of two classes. Table 3 illustrates their characteristics in terms of number of samples, number of input features, type of features and source. We have four types of inputs: boolean; categorical; integer; and real. The public source of the datasets is https://archive.ics.uci.edu/ml/datasets.html. Our experiments are based on 10 repetitions of 10-fold cross-validation trials. Training sets were normalized using Gaussian normalization. We compared three variants of our approach: (1) App1_DIMLP that uses the electic rule extraction algorithm described in [7]; (2) App2_TREPAN that uses the pedagogical rule extraction algorithm described in [12]; (3) App3_Extract that replaces the DMLPs and the rule extraction step by a rule extraction algorithm that extracts rules directly from the bootstrap samples. In order to validate the performance of our approach, we compared the three variants described above to: the most popular ensemble learning methods (Bagging [10], AdaBoost [18]) and two classification approaches based on ensemble rule extraction that uses the DIMPL [8]: one trained by bagging (DIMLP-B) and another trained by arcing (DIMLP-A).

Table 3. Datasets properties

Dataset	Number of input features	Number of samples	Type of features
Breast cancer prognastic	33	194	real
Glass	9	163	real
Haberman	3	306	int
Heart disease	13	270	bool, cat, int, real
ILPD (Liver)	10	583	int, real
Pima Indians	8	768	int, real
Saheart	9	462	bool, int, real
Sonar	60	280	Real
Spect heart	22	267	bin
Vertebral column	6	310	real

We defined a grid search to optimize the parameters of each approach. The number of the bootstrap samples used in all the approaches is shown in Table 4. For DIMLP ensembles, we have used the default parameters defined in [8] (for example, the number of bootstrap samples is equal to 25). In the experiment, we have used Accuracy and Fidelity as the evaluation measures to compare the classification performance of different methods described above. Accuracy indicates the percentage of well-predicted data and Fidelity indicates the degree of matching between network classifications and rules' classifications.

Table 4. Results comparison to ensemble methods.

Datasets	Adaboost Accuracy	Bagging Accuracy	DIMLP-B Accuracy	DIMLP-A Accuracy	App3_Extract Accuracy	App1_DIMLP Accuracy	Fidelity	App2_TREPAN Accuracy	Fidelity
Breast Cancer Prognastic	81±0.01 (150)	77.3±0.02 (125)	79.0±0.08 (25)	77.7±0.04 (25)	79.1±0.11 (24)	**81.2**±0.03 (22)	96.6±0.02	79.4±0.05 (24)	95.9±0.04
Glass	80.3±0.09 (100)	81.0±0.06 (100)	77.8±0.06 (25)	**81.1**±0.02 (25)	75.8±0.08 (25)	78.6±0.10 (21)	95.5±0.10	74.6±0.09 (25)	94.8±0.10
Haberman	69.0±0.16 (100)	71.3±0.11 (125)	74.3±0.02 (25)	73.3±0.06 (25)	69.9±0.01 (25)	76.3±0.22 (25)	95.8±0.01	**77.2**±0.11 (24)	93.9±0.06
Heart Disease	86.0±0.09 (100)	85.9±0.02 (100)	84.3±0.06 (25)	80.5±0.12 (25)	82.7±0.11 (26)	**86.6**±0.01 (25)	97.1±0.01	76.9±0.13 (21)	96.8±0.07
ILPD (Liver)	72.5±0.08 (150)	71.0±0.09 (125)	70.7±0.21 (25)	70.8±0.11 (25)	69.9±0.06 (25)	**73.7**±0.20 (21)	96.8±0.08	69.7±0.07 (26)	95.1±0.03
Pima Indians	**77.0**±0.02 (100)	76.0±0.06 (100)	76.3±0.05 (25)	74.2±0.09 (25)	71.2±0.08 (22)	76.9±0.02 (23)	96.8±0.06	75.6±0.06 (22)	96.1±0.08
Saheart	71.7±0.16 (150)	71.0±0.10 (100)	71.9±0.21 (25)	68.6±0.06 (25)	69.9±0.02 (25)	**72.9**±0.02 (21)	96.9±0.01	71.9±0.13 (25)	93.8±0.01
Sonar	71.0±0.20 (100)	77.6±0.11 (100)	79.0±0.10 (25)	78.4±0.09 (25)	71.9±0.06 (21)	**79.8**±0.09 (25)	95.9±0.01	76.9±0.16 (23)	94.7±0.08
Spect Heart	66.9±0.13 (125)	71.2±0.05 (150)	72.2±0.11 (25)	67.9±0.22 (25)	61.9±0.22 (25)	**72.4**±0.02 (24)	95.9±0.05	71.9±0.08 (25)	96.4±0.05
Vertebral Column	67.3±0.13 (125)	72.3±0.06 (150)	84.0±0.03 (25)	82.7±0.05 (25)	81.8±0.06 (25)	**86.6**±0.06 (25)	94.9±0.12	70.2±0.03 (22)	96.9±0.02

From the experimental results shown in Table 4, we can see that our approach can effectively ensure high accuracy in classification. We can see that, App1_DIMLP and App2_TREPAN as ensemble learning methods give better results than Bagging and AdaBoost methods. For example in Vertebral Column dataset, App1_DIMLP obtains an accuracy of up to 86.6% (using 25 classifiers DIMLPs) while Bagging and AdaBoost are lower than 73%. This can be explained by the use of the argumentation, a novel strategy for classifiers combination, which is more transparent than the combination of classifiers used in usual ensemble methods (such as voting). Our method can outperform DIMLP-B and DIMLP-A on the majority of datasets. For example, in heart disease dataset, the accuracy of App1_DIMLP is higher than that of DIMLP-A by 6%. In Breast Cancer Prognastic dataset, the accuracy of App1_DIMLP is 81.2% (using 22 classifiers DIMLPs) however the accuracy of DIMLP-B and DIMLP-A are lower than 80%. So far, the results was in our favour for predictive accuracy in 9 out of 10 classification problems. Moreover The Fidelity is higher than 93% in all datasets. This means that the network classifications match rules classifications.

App1_DIMLP and App2_TREPAN produce better results than App3_Extract. This can be explained by the power of prediction of DMLP. Indeed, the rule extraction from DMLP allows to ensure higher classification accuracy than a direct rules extraction from the bootstrap samples. As a conclusion, from the above experimental results, we can find that App1_DIMLP and App2_TREPAN can effectively extract high quality knowledge for ensemble classifier and ensure high accuracy in classification as well, which indicates that argumentation as a novel ensemble strategy can improve the capability of knowledge integration effectively. Moreover our method provides transparency of the predictions since it can provide an intelligible explanation and extract useful knowledge from ensemble classifiers.

6   Conclusion

In order to improve performance classification, we have proposed a transparent deep ensemble method based argumentation for classification. In this method, argumentation is used as an ensemble strategy for deep learning algorithms combination, which is more comprehensible and explicable than traditional ensemble method (such as voting). Meanwhile, by using argumentation, we improved performance classification. Experiments show that, as ensemble method, our approach significantly outperforms single classifiers and traditional ensemble methods. In addition, our method effectively provides explanation behind decisions and therefore addresses the recent need for Explainable AI. The explanation provided to the user is easy to grasp so he will be able to judge the acceptance of decisions. Moreover, other agents containing rules about domain knowledge can be easily added. The prospects of this work are various. In the short term, it will be necessary to carry out experiments on a larger scale to consolidate the results of our approach with the real electronic health records in order to realize several tasks such as dignosis, the prediction of the next visit date, etc. We also plan to apply our approach to other types of deep learning algorithms such as convolutional neural networks, recurrent neural networks, etc. To improve the interaction process, we will use more complex negotiation rather than a limited Agent-to-agent exchange based on the speech acts.

References

1. Adadi, A., Berrada, M.: Peeking inside the black-box: a survey on explainable artificial intelligence (XAI). IEEE Access **6**, 52138–52160 (2018)
2. Agrawal, R., Imieliński, T., Swami, A.: Mining association rules between sets of items in large databases. SIGMOD Rec. **22**(2), 207–216 (1993). https://doi.org/10.1145/170036.170072
3. Amgoud, L., Parsons, S., Maudet, N.: Arguments, dialogue, and negotiation. In: ECAI (2000)
4. Andrews, R., Diederich, J., Tickle, A.B.: Survey and critique of techniques for extracting rules from trained artificial neural networks. Knowl.-Based Syst. **8**(6), 373–389 (1995). https://doi.org/10.1016/0950-7051(96)81920-4
5. Augasta, M.G., Kathirvalavakumar, T.: Reverse engineering the neural networks for rule extraction in classification problems. Neural Process. Lett. **35**(2), 131–150 (2012). https://doi.org/10.1007/s11063-011-9207-8
6. Besnard, P., et al.: Introduction to structured argumentation. Argument Comput. **5**(1), 1–4 (2014). https://doi.org/10.1080/19462166.2013.869764
7. Bologna, G., Hayashi, Y.: A rule extraction study on a neural network trained by deep learning. In: 2016 International Joint Conference on Neural Networks, IJCNN 2016, Vancouver, BC, Canada, 24–29 July 2016, pp. 668–675. IEEE (2016). https://doi.org/10.1109/IJCNN.2016.7727264
8. Bologna, G., Hayashi, Y.: A comparison study on rule extraction from neural network ensembles, boosted shallow trees, and SVMs. Appl. Comput. Intell. Soft Comput. **2018**, 1–20 (2018)

9. Bonzon, E., Maudet, N.: On the outcomes of multiparty persuasion. In: McBurney, P., Parsons, S., Rahwan, I. (eds.) ArgMAS 2011. LNCS (LNAI), vol. 7543, pp. 86–101. Springer, Heidelberg (2012). https://doi.org/10.1007/978-3-642-33152-7_6

10. Breiman, L.: Bagging predictors. Mach. Learn. **24**, 123–140 (1996)

11. Cocarascu, O., Toni, F.: Detecting deceptive reviews using argumentation. In: Proceedings of the 1st International Workshop on AI for Privacy and Security, PrAISe 2016, pp. 9:1–9:8. ACM, New York (2016). https://doi.org/10.1145/2970030.2970031

12. Craven, M., Shavlik, J.W.: Using sampling and queries to extract rules from trained neural networks. In: ICML (1994)

13. Craven, M.W., Shavlik, J.W.: Extracting tree-structured representations of trained networks. In: Proceedings of the 8th International Conference on Neural Information Processing Systems, NIPS 1995, pp. 24–30. MIT Press, Cambridge (1995)

14. Dietterich, T.G.: Ensemble methods in machine learning. In: Kittler, J., Roli, F. (eds.) MCS 2000. LNCS, vol. 1857, pp. 1–15. Springer, Heidelberg (2000). https://doi.org/10.1007/3-540-45014-9_1

15. Dietterich, T.G., Bakiri, G.: Solving multiclass learning problems via error-correcting output codes. J. Artif. Intell. Res. **2**, 263–286 (1995). http://dblp.uni-trier.de/db/journals/jair/jair2.html#DietterichB95

16. Dung, P.M.: On the acceptability of arguments and its fundamental role in non-monotonic reasoning, logic programming and n-person games. Artif. Intell. **77**(2), 321–357 (1995)

17. Forgy, C.: Rete: a fast algorithm for the many pattern/many object pattern match problem. Artif. Intell. **19**(1), 17–37 (1982)

18. Freund, Y., Schapire, R.E.: Experiments with a new boosting algorithm (1996)

19. Fu, L.: Rule generation from neural networks. IEEE Trans. Syst. Man Cybern. **24**, 1114–1124 (1994)

20. Garcia, F.J.C., Robb, D.A., Liu, X., Laskov, A., Patrón, P., Hastie, H.F.: Explain yourself: a natural language interface for scrutable autonomous robots. CoRR arXiv:abs/1803.02088 (2018)

21. Guidotti, R., Monreale, A., Ruggieri, S., Turini, F., Giannotti, F., Pedreschi, D.: A survey of methods for explaining black box models. ACM Comput. Surv. **51**(5), 93:1–93:42 (2018)

22. Prakken, H.: Models of persuasion dialogue. In: Simari, G., Rahwan, I. (eds.) Argumentation in Artificial Intelligence, pp. 281–300. Springer, Boston (2009). https://doi.org/10.1007/978-0-387-98197-0_14

23. Hao, Z., Yao, L., Liu, B., Wang, Y.: Arguing prism: an argumentation based approach for collaborative classification in distributed environments. In: Decker, H., Lhotská, L., Link, S., Spies, M., Wagner, R.R. (eds.) DEXA 2014. LNCS, vol. 8645, pp. 34–41. Springer, Cham (2014). https://doi.org/10.1007/978-3-319-10085-2_3

24. Harbers, M.: Self-explaining agents in virtual training. In: EC-TEL PROLEAN (2008)

25. Hruschka, E.R., Ebecken, N.F.: Extracting rules from multilayer perceptrons in classification problems: a clustering-based approach. Neurocomputing **70**(1), 384–397 (2006). https://doi.org/10.1016/j.neucom.2005.12.127. http://www.sciencedirect.com/science/article/pii/S0925231206000403, Neural Networks

26. Johnson, W.L.: Agents that learn to explain themselves. In: Proceedings of the Twelfth National Conference on Artificial Intelligence, AAAI 1994, vol. 2, pp. 1257–1263. American Association for Artificial Intelligence, Menlo Park (1994)

27. Kingma, D.P., Ba, J.: Adam: a method for stochastic optimization. arXiv preprint arXiv:1412.6980 (2014)

28. Kontarinis, D.: Debate in a multi-agent system: multiparty argumentation protocols (2014)
29. van Lent, M., Fisher, W., Mancuso, M.: An explainable artificial intelligence system for small-unit tactical behavior. In: Proceedings of the 16th Conference on Innovative Applications of Artificial Intelligence, IAAI 2004, pp. 900–907. AAAI Press (2004)
30. Lipton, Z.C.: The mythos of model interpretability. Queue **16**(3), 30:31–30:57 (2018). https://doi.org/10.1145/3236386.3241340
31. Lu, H., Setiono, R., Liu, H.: Effective data mining using neural networks. IEEE Trans. Knowl. Data Eng. **8**(6), 957–961 (1996). https://doi.org/10.1109/69.553163
32. Sato, M., Tsukimoto, H.: Rule extraction from neural networks via decision tree induction. In: International Joint Conference on Neural Networks (IJCNN 2001), pp. 1870–1875 (2001)
33. Marchant, I., et al.: Score should be preferred to Framingham to predict cardiovascular death in French population. Eur. J. Cardiovasc. Prev. Rehabil. **16**, 609–615 (2009)
34. Mcburney, P., Parsons, S.: Dialogue games in multi-agent systems. Informal Logic **22**, 2002 (2002)
35. Molineaux, M., Dannenhauer, D., Aha, D.W.: Towards explainable NPCS: a relational exploration learning agent. In: AAAI Workshops (2018)
36. Re, M., Valentini, G.: Ensemble methods: a review, pp. 563–594 (2012)
37. Reed, C.: Representing dialogic argumentation. Knowl.-Based Syst. **19**, 22–31 (2006)
38. Sato, M., Tsukimoto, H.: Rule extraction from neural networks via decision tree induction. In: IJCNN 2001, vol. 3, pp. 1870–1875 (2001)
39. Searle, J.: Speech Acts. An Essay in the Philosophy of Language. Cambridge University Press, Cambridge (1969)
40. Thimm, M., Kersting, K.: Towards argumentation-based classification. In: Logical Foundations of Uncertainty and Machine Learning, Workshop at IJCAI 2017, August 2017. http://www.mthimm.de/publications.php
41. Tran, S.N., d'Avila Garcez, A.: Knowledge extraction from deep belief networks for images. In: IJCAI 2013 Workshop on Neural-Symbolic Learning and Reasoning (2013)
42. Wardeh, M., Bench-Capon, T., Coenen, F.: Arguing from experience using multiple groups of agents. Argument Comput. **2**(1), 51–76 (2011)
43. Wardeh, M., Coenen, F., Bench-Capon, T.: Multi-agent based classification using argumentation from experience. Auton. Agents Multi-Agent Syst. **25**(3), 447–474 (2012). https://doi.org/10.1007/s10458-012-9197-6
44. Zilke, J.R., Loza Mencía, E., Janssen, F.: DeepRED – rule extraction from deep neural networks. In: Calders, T., Ceci, M., Malerba, D. (eds.) DS 2016. LNCS (LNAI), vol. 9956, pp. 457–473. Springer, Cham (2016). https://doi.org/10.1007/978-3-319-46307-0_29

Effects of Agents' Transparency on Teamwork

Silvia Tulli$^{(\boxtimes)}$ ⓘ, Filipa Correia ⓘ, Samuel Mascarenhas ⓘ, Samuel Gomes ⓘ,
Francisco S. Melo ⓘ, and Ana Paiva ⓘ

Department of Computer Science and Engineering,
INESC-ID and Instituto Superior Técnico, Universidade de Lisboa,
2744-016 Porto Salvo, Portugal
{silvia.tulli,samuel.mascarenhas}@gaips.inesc-id.pt,
{filipacorreia,samuel.gomes}@tecnico.ulisboa.pt,
{francisco.melo,ana.paiva}@inesc-id.pt

Abstract. Transparency in the field of human-machine interaction and artificial intelligence has seen a growth of interest in the past few years. Nonetheless, there are still few experimental studies on how transparency affects teamwork, in particular in collaborative situations where the strategies of others, including agents, may seem obscure.

We explored this problem using a collaborative game scenario with a mixed human-agent team. We investigated the role of transparency in the agents' decisions, by having agents that reveal and tell the strategies they adopt in the game, in a manner that makes their decisions transparent to the other team members. The game embraces a social dilemma where a human player can choose to contribute to the goal of the team (cooperate) or act selfishly in the interest of his or her individual goal (defect). We designed a between-subjects experimental study, with different conditions, manipulating the transparency in a team. The results showed an interaction effect between the agents' strategy and transparency on trust, group identification and human-likeness. Our results suggest that transparency has a positive effect in terms of people's perception of trust, group identification and human likeness when the agents use a tit-for-tat or a more individualistic strategy. In fact, adding transparent behaviour to an unconditional cooperator negatively affects the measured dimensions.

Keywords: Transparency · Autonomous agents ·
Multi-agent systems · Public goods game · Social dilemma

This work was supported by national funds through Fundação para a Ciência e a Tecnologia (FCT-UID/CEC/50021/2019), and Silvia Tulli acknowledges the European Union's Horizon 2020 research and innovation program for grant agreement No. 765955 ANIMATAS project. Filipa Correia also acknowledges an FCT grant (Ref. SFRH/BD/118031/2016).

D. Calvaresi et al. (Eds.): EXTRAAMAS 2019, LNAI 11763, pp. 22–37, 2019.
https://doi.org/10.1007/978-3-030-30391-4_2

1   Introduction

The increase of intelligent autonomous systems capable of complex decision-making processes affects humans' understanding of the motivations behind the system's responses [6]. In this context, evaluating the performance of machine learning algorithms may not be sufficient to prove the trustworthiness and reliability of a system in the wild [25].

Machine learning models appear to be opaque, less intuitive and challenging for the diversified end users. To meet this need, an increasing number of studies has focused on developing transparent systems. However, the definition of transparency is still up for debate. The most commonly used terms are model interpretability, explicability, reliability, and simplicity. Doshi-Velez and Kim define interpretability as the ability to explain or present understandable terms to a human [12]. Instead, Rader et al. explain transparency as providing the non-obvious information that is difficult for an individual to learn or experience directly, such as how and why a system works the way it does and what its outputs mean [26]. The lack of a consensual definition of transparency reflects in a lack of comparable metrics to assess it. Due to this, to understand transparency, it is necessary to manipulate and measure various factors that can influence the perception and behavior of humans. Designing the transparency of a system is therefore not a purely computational problem.

A variety of human challenges demands for effective teamwork [18]. However teamwork has numerous implications: the commitment of all the members to achieve the team goals, the trust among the team members, the mutual predictability for effective coordination, and the capability to adapt to changing situations [19,22]. Many of the features needed for successful teamwork are well illustrated in video games scenarios [14], and due to this, video games have become a popular object of investigation for social and cultural sciences [23]. When autonomous systems move from being tools to being teammates, an expansion of the model is needed to support the paradigms of teamwork, which require two-way transparency [6]. As in human-human groups, the communication of relevant information can facilitate analysis and decision-making by helping the creation of a shared mental model between the group members. Several studies based on human-agent collaboration suggest that humans benefit from the transparency of the agent, which consequently improves the cooperation between them [26]. Moreover, agents' transparency facilitates the understanding of the responsibilities that different group members might take in collaborative tasks.

Contrary to what we could hypothesize, collaborative games can also encourage anti-collaborative practices that derive from the identification of a single winner and from the fact that players rely on the contribution of others and therefore invest less in their actions (free riding) [4]. For this reason, combining the investigation of the behavioral model of the players in relation to the different strategies of the team members and the transparency of the decision-making process of the artificial players turns out to be useful for the design of systems that aim to facilitate and foster collaboration. The objective of this study is to investigate the effect of the transparency and strategy of virtual agents on human pro-social behavior in a collaborative game.

2   Related Work

The lack of transparency is considered one of the obstacles for humans to establish trust towards autonomous systems [10]. In fact, trust appears as a common measure to assess the effect of transparency and it is related to the level of observability, predictability, adjustability, and controllability, as well as mutual recognition of common objectives of a system. Chen et al. have developed a model for collaboration and mutual awareness between humans and agents [6]. This model is called *Situation Awareness Based Agent Transparency* (SAT) and considers current plans and actions, decision-making and prediction of responses. To sum up, the SAT model describes the type of information that the agent should provide on its decision-making process to facilitate mutual understanding and collaboration between human and agent. The first level of the model includes information related to the actions, plans, and objectives of the agent. This level helps human's perception of the current state of the agent. The second level considers the decision-making process with the constraints and affordances that the agent takes into account when planning its actions. With that, the human can understand the current behavior of the agent. The third level provides information related to the agent's projection towards future states with the relative possible consequences, the probability of success or failure, and any uncertainty associated with the previously mentioned projections. The third level allows the human to understand the future responses of the agent. Our manipulation of the agents' transparency considers the three levels of the SAT informing about the current actions and plans, and including the decision-making process (e.g. "My plan is to always improve the instrument"). The third level results as a projection of the pursued strategy.

Given that, it can be difficult to distinguish in the literature whether transparency refers to the mechanism or the outcome, the cause or the effect [26]. However, in the context of human-machine interaction, transparency means an appropriate mutual understanding and trust that leads to effective collaboration between humans and agents. The act of collaboration and cooperation in group interactions is not only interesting for researchers in the area of human-machine interaction but is also widely studied by social sciences to obtain knowledge on how cooperation can be manipulated. In particular, to understand how individuals in a group can be stimulated to contribute to a public good [13]. Several studies, both theoretically and empirically, shown that transparency has a positive effect on cooperation. For instance, Fudenberg et al. demonstrated that transparency of past choices by the group members is necessary to maintain a sustainable and stable cooperation [15]. Davis et al. shown that transparency allows cooperative players to indicate their cooperative intentions, which may induce others to similar cooperative behaviors [11].

3   Research Design and Methods

We conducted a between-subject user study using the Mechanical Turk and the "For The Record" game [9]. "For the Record" is a public goods game that embraces a social dilemma where a human player can choose to contribute to the goal of the team (cooperate) or act selfishly in the interest of his or her individual goal (defect). In linear public goods environments *maximizers have a dominant strategy to either contribute all of their tokens or none of their tokens to a group activity* [5,28]. In the "For The Record" experimental scenario, three players, one human, and two artificial agents, have the goal of publishing as many albums as possible. The number of albums to be created and produced matches the number of rounds to play, in our case, 5 rounds and if players fail 3 albums they lose the game. During the first round, each player starts playing by choosing the preferred instrument that can be used to create the album. Starting from the second round each player has two possible actions and they concern the possibility of investing in the instrument's ability (contributing to the success of the album) or in the marketing's ability (contributing to the individual monetary value, or personal profit, obtained after the album's success). This investment is translated into the number of dice that the player can use, in the first case to play the instrument and helps to create the album, while in the second case to receive profit. During the creation of the album, each player will contribute equally to the value obtained from the roll of the dice, and the number of die available to the player will depend on the level/value of the skill (marketing or instrument). The score of an album consists of adding up the values achieved by each player during his performance. After creating the album, the band has to release it on the market. The market value is evaluated by rolling 2 dice of 20 faces. If the market value is higher than the album score, than the album is considered a "Fail". On the other hand, if the market value is less than or equal to the score on the album, that album is considered a "Mega-hit". From the fourth round on, the band enters the international market, which means that the market value is evaluated by rolling 3 die of 20 faces (instead of the 2 previous dices). This increases the difficulty of getting successful albums. The game has always been manipulated to return a victory.

4   Objective and Hypothesis

The objective of this study was to investigate the effect of the transparency and strategy of virtual agents on human pro-social behavior in a collaborative game. Despite having hypothesized that transparency would affect several measures of teamwork, we have also manipulated the agents' strategy to confirm if the results would provide similarly when the agents adopted different strategies. In a two by three (2×3) between-subjects design, resulting in six experimental conditions, we manipulated the agents' transparency and the agents' strategy, respectively. The two levels of transparency were:

- **Transparent:** The agents explain their strategy;
- **Non-transparent:** The agents do not explain their strategy.

The three possible strategies for the agents were:

- **Cooperative:** The agents always cooperate;
- **Individualistic:** The agents cooperate only if the last round has been lost;
- **Tit for Tat:** The agents cooperate only if the player cooperate.

We expected that the transparency of the agents will positively affect teamwork and make the agents' strategy easily to interpret. We also expected transparency to increase trust and facilitate collaboration due to mutual understanding and shared responsibilities. Therefore we have the following hypotheses:

- H1: The agents' transparency increases the number of cooperative choices of the human player;
- H2: The agents' transparency results in greater trust and group identification;
- H3: The agents' transparency increases the likeability and human likeness of the artificial player;

The hypothesis that the transparency increases the number of cooperative choices is based on the fact that transparency about choices tends to lead to an increase in contributions and collusion [13]. The hypothesis that positive effect of transparency on trust and group identification relies on the evidence that transparency have the (perhaps counter-intuitive) quality of improving operators' trust in less reliable autonomy. Revealing situations where the agent has high levels of uncertainty develops trust in the ability of the agent to know its limitations [7, 8, 16, 24]. The hypothesis that the agents' transparency results in greater likeability and perceived human likeness of the artificial player refers to the experimental evidence of Herlocker et al. showing that explanations can improve the acceptance of automated collaborative filtering (ACF) systems [17].

4.1   Materials and Methods

Agents' Transparency Manipulation. The interactive agents commented some game events through text in speech bubbles, e.g., *That was very lucky!* or *Lets record a new album.*

The duration of such stimuli depend on the number of words shown, according to the average reading speed of 200–250 words per minute. However, the speech bubbles containing the manipulation of each experimental condition lasted twice as much to make sure the participants would read them (Fig. 1).

Table 1 shows the explanation given by the artificial agents while they are choosing the main action of adding a point to either the instrument or the marketing in the transparent and non-transparent conditions:

In the non-transparent conditions the agents explain what they are doing for that current round, in the transparent conditions they explicitly refer to their plans and intentions.

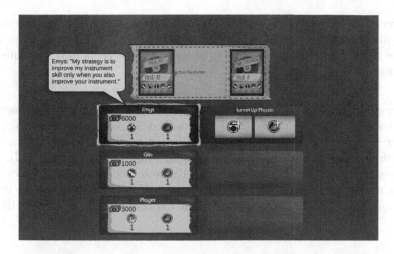

Fig. 1. Example of a speech bubble with the explanation of the agents' strategy

Table 1. Manipulation of transparent and non-transparent behaviour for each agents' strategy

Strategy	Transparency	
	Transparent	Non-transparent
Cooperative	1. *"My strategy is to always improve the instrument."* 2. *"My plan is to always improve the instrument."*	1. *"I am going improve the [instrument/marketing]."* 2. *"I will put one more point on my [instrument/marketing]."*
Individualistic	1. *"My plan is to improve my marketing skill only when the album success."* 2. *"My plan is to improve my instrument skill only when the album fails."*	
Tit for tat	1. *"My strategy is to improve my instrument skill only when you also improve your instrument."* 2. *"My strategy is to improve my marketing skill only when you also improve your marketing."*	

4.2 Metrics and Data Collection

To test our hypotheses and, therefore, analyse the effects of the strategy and transparency adopted by the agents, we used different metrics and items from standardized questionnaires. The self-assessed questionnaire included some demographic questions (e.g., age, gender and ethnicity), a single-item on their self-perceived competitiveness level, two items regarding the naturalness and human-likeness of the agents' strategies, and two validation questions to evaluate the understanding on the rules of the game. The remaining measures are detailed as follows.

Cooperation Rate. The cooperation rate was an objective measure assessed during the game-play. In the beginning of each round, each player has to choose between to cooperate with the team (i.e., by upgrading the instrument skill) or to defect for individual profit (i.e., by upgrading the marketing skill). This measure sums up the total number of times the human player opted to cooperate and can range, in discrete numbers, from zero to four. It represents the degree of pro-sociality that the human participant expressed while teaming with the agents.

Group Trust. We chose the Trust items by Allen et al. in [1], which were explicitly designed for virtual collaboration to assess the trust through the agents. Trust is described as a key element of collaboration and is divided into seven items with a 7 points likert-scale from totally disagree to totally agree.

Multi-component Group Identification. Leach et al. identified a set of items for the assessment of the Group-Level Self-Definition and Self-Investment in [21]. The idea behind this scale is that individuals' membership in groups has relevant impact on humans behavior. Specifically designed items represents the five dimensions evaluated: individual self-stereotyping, in-group homogeneity, solidarity, satisfaction, and centrality. These items were presented with a Likert-type response scale that ranged from 1 (strongly disagree) to 7 (strongly agree). We decided to use the dimensions of homogeneity, solidarity and satisfaction as relevant metrics for our study.

Godspeed. The Godspeed scale was designed for evaluating the perception of key attributes in Human-Robot Interaction [3]. More precisely, the scale is meant to measure the level of anthropomorphism, animacy, likeability, perceived intelligence, and perceived safety. Each dimension has five or six items with semantic differentials couples that respondents are asked to evaluate in a 5 points Likert scale. We used the dimensions of the likeability (Dislike/Like, Unfriendly/Friendly, Unkind/Kind, Unpleasant/Pleasant, Awful/Nice) and perceived intelligence (Incompetent/Competent, Ignorant/Knowledgeable, Irresponsible/Responsable, Unintelligent/Intelligent, Foolish/Sensible).

4.3   Procedure

Participants were asked to complete the task in around 40 min. The experiment was divided in three phases. The first phase consisted of the game tutorial, and lasted around 15 min. The second phase consists in playing a session of

"For the Record" with the two artificial agents, which lasted around 15 min. The last phase was represented by the questionnaire and took round 10 min. We informed the participants about the confidentiality of the data, voluntary participation and the authorization for sharing the results with the purpose of analysis, research and dissemination. We specified that we were interested in how people perceive teamwork and the game strategies of the two artificial players they were going to play with. After finishing the experiment and providing their judgments, we thanked the participants for their participation giving them 4$. We collected the data for the non transparent and the transparent condition separately, ensuring that none of the participants repeat the experiment twice.

5 User Study

The main goal of our study was to explore the role transparent behaviors have on the perception of intelligent agents during human-agent teamwork. In particular, to analyze if transparency can enhance the perception of the team and the display of pro-social behaviors by humans.

5.1 Participants

The participants involved in the study were 120, 20 participants per each experimental condition (Cooperative, Individualistic and Tit for Tat). Considering the study was done in MTurk and the fact that the experiment took more time than the turkers are used to, we introduced some attention and verification questions in order to ensure the quality of the data. The criteria to exclude participants were: not having completed the entire experiment; having reported an incorrect score of the game; and having provided wrong answers to the questions related to the game rules (e.g., *How many dices are rolled for the international market?*). Consequently, we run the data analysis on a sample of a sample of 80, 28 in the non-transparency conditions and 52 in the transparency conditions. The average age of the sample was 37 years (min = 22, max = 63, stdev = 8.78) and was composed of 52 males and 27 females and one other. The participants were randomly assigned to one of three condition of the strategy: 19 in the cooperative condition (13 in the transparency condition and 6 in the non-transparency condition), 30 in the individualistic condition (17 for the transparency condition and 13 in the non-transparency condition), 18 for the tit-for-tat condition (9 for the transparency condition and 11 in the non-transparency condition).

5.2 Data Analysis

We analyzed the effects of our independent variables - transparency (binary categorical variable *Transparent* and *Non-Transparent*) and strategy (three categories: *Cooperative*, *individualistic* and *Tit for Tat*) - on the dependent variables.

The reliability analysis for the dimensions of the Trust scale, the Group Identification scale, the Godspeed scale as well as the Human likeness and Naturalness revealed excellent internal consistency among items of the same dimensions (Trust: $\alpha = 0.912$; Group Identification: $\alpha = 0.972$; Group Solidarity: $\alpha = 0.953$; Group Satisfaction: $\alpha = 0.969$; Group Homogeneity: $\alpha = 0.923$; Perceived Intelligence: $\alpha = 0.962$; Likeability: $\alpha = 0.978$; Human-likeness and Naturalness: $\alpha = 0.938$).

Cooperative Rate. The analysis of the number of defects, revealed that the main effect of transparency was not significant ($F(1, 73) = 0.320, p = 0.573$), and the main effect of strategy was not significant ($F(3, 73) = 2.425, p = 0.072$). The interaction effect between the two factors was not significant ($F(2, 73) = 0.003, p = 0.997$). The specific values per each strategy were: Cooperative (M = 1.11, SE = 0.201, SD = 875), Individualistic (M = 1.70, SE = 0.153, SD = 0.837), Tit for Tat (M = 1.06, SE = 0.249, SD = 1.056).

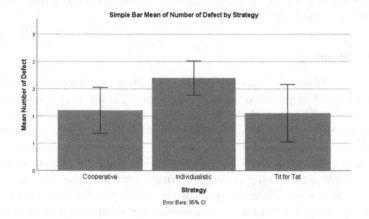

Fig. 2. Number of defects by strategy

Group Trust. The Analysis of Variance in Trust, showed that the main effect of transparency was not significant (F(1,73) = 0.337, p = 0.563), and the main effect of strategy was significant ($F(3, 73) = 8.117, p < 0.001$). The specific values for each strategy were: Cooperative (M = 5.25, SE = 0.265, SD = 1.154), Individualistic (M = 4.42, SE = 0.230, SD = 1.261), Tit for Tat (M = 5.22, SE = 0.221, SD = 0.938).

The interaction effect between the two factors was significant ($F(2, 73) = 3.833, p = 0.026$).

Figure 3 shows that only in the cooperative condition the transparency negatively influenced the level of trust towards the agents. The specific values per each strategy in the transparent and non-transparent conditions were: Transparent - Cooperative (M = 4.90, SE = 0.334, SD = 1.204), individualistic (M = 4.89,

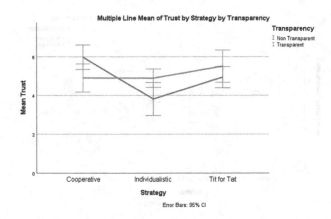

Fig. 3. Interaction effect between strategy and transparency in trust

SE $= 0.224$, SD $= 0.925$), Tit for Tat (M $= 5.51$, SE $= 0.362$, SD $= 1.086$) Non-Transparent - Cooperative (M $= 5.98$, SE $= 0.246$, SD $= 0.602$), Individualistic (M $= 3.81$, SE $= 0.291$, SD $= 1.411$), Tit for Tat (M $= 4.95$, SE $= 0.239$, SD $= 0.711$).

Multi-component Group Identification. The Group Identification, did not reveal main effect of single factors of transparency and strategy ($F(1, 73) = 2.674; F(3, 73) = 2.360, p = 0.106, p = 0.078$). However, the interaction between the two factors was significant ($F(2, 73) = 4.320, p = 0.017$). The specific values per each strategy in the transparent and non-transparent conditions: Transparent - Cooperative (M $= 4.15$, SE $= 0.500$, SD $= 1.801$), Individualistic (M $= 5.06$, SE $= 0.336$, SD $= 1.387$), Tit for Tat (M $= 5.27$, SE $= 0.427$, SD $= 1.282$). Non-Transparent - Cooperative (M $= 5.19$, SE $= 0.394$, SD $= 0.965$), Individualistic (M $= 3.32$, SE $= 0.359$, SD $= 1.292$), Tit for Tat (M $= 3.98$, SE $= 0.559$, SD $= 1.676$).

As we can notice from the Fig. 4, transparency and strategy influenced the perception of Group Identification in the opposite direction among the agents' strategies. In the transparency condition, the agents foster less group identification when they acts cooperatively. However, transparency had a positive influence in the group identification in the Individualistic and Tit for Tat condition. The One-way ANOVA in Group Identification reveals that the effect of transparency in Cooperative condition was not significant ($F(1, 17) = 1.732, p = 0.206$), the effect of transparency in Individualistic condition was significant ($F(1, 28) = 12.178, p = 0.002$) and the effect of transparency in Tit for Tat condition was not significant ($F(1, 16) = 3.398, p = 0.084$).

Goodspeed. The Likeability did not reveal a main effect of transparency ($F(1, 73) = 0.001, p = 0.973$) but informed a main effect of the strategy on the likeability ($F(3, 73) = 3.279, p = 0.026$) Fig. 5. The interaction between the

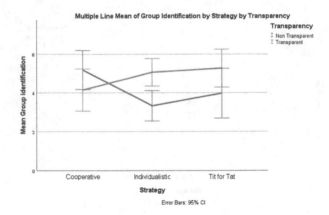

Fig. 4. Interaction effect between strategy and transparency in Group Identification

transparency and strategy was not significant $(F(2,73) = 0.855, p = 0.429)$. Again in this case, the strategy affected the perception of likeability, and no interaction was found regardless of whether or not the agents employ transparent behaviors.

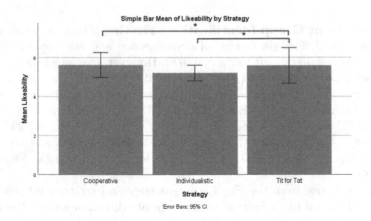

Fig. 5. Main effect of the strategy on likeability

For the human-likeness dimension, there was no main effect of transparency $(F(1,73) = 0.145, p = 0.704)$ and no main effects of the strategy $(F(3,73) = 2.181, p = 0.098)$. However, there was a significant interaction effect between transparency and strategy for the Human-likeness attributed to the agents $(F(2,73) = 3.585, p = 0.033)$.

In Fig. 6 we confirmed the trend of a different effect of transparency in the cooperative condition in respect to the strategy. For the Tit for Tat condition we can notice that both strategy and transparency positively affect the perceived human likeness of the agents.

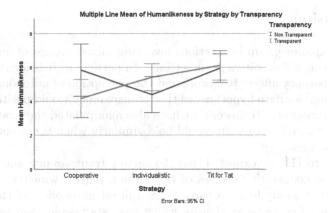

Fig. 6. Interaction effect between strategy and transparency in humanlikeness

The Univariate Analysis of Variance of the transparency and strategy for the Perceived Intelligence informed that the main effect of transparency was not significant $(F(1, 73) = 0.652, p = 0.422)$, but the main effect of strategy was significant $(F(3, 73) = 5.297, p = 0.002)$ Fig. 7. The interaction effect between the two fixed factors was not significant $((2, 73) = 3.632, p = 0.179)$. In other words, only the strategy of the agents, regardless of whether or not the agents employ transparent behaviors, affects the perceived intelligence of the agents, in particular for the Tit for Tat strategy as confirmed by several studies about game theory [2, 27]. The specific values per each strategy were: Cooperative (M = 5.39, SE = 0.348, SD = 1.518), Individualistic (M = 5.23, SE = 0.227, SD = 1.244), Tit for Tat (M = 6.11, SE = 0.249, SD = 1.054).

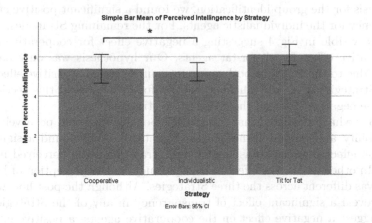

Fig. 7. Main effect of the strategy on perceived intelligence

6   Discussion

This paper explores group interactions involving mixed groups of humans and virtual agents in collaborative game settings. In particular, it is focused on how agents' transparency affects teamwork and the perception of autonomous teammates. Although we have hypothesized that transparency would positively influence several measures of teamwork, we have also manipulated the strategy of the agents to ascertain if the results would hold similarly when the agents adopted different Strategies.

According to **H1**, we expected that the agents' transparency would increase the number of cooperative choices of the human player, which was not confirmed. In fact, we only found a partially significant main effect of the strategy on the number of cooperative choices, which suggests people cooperated differently according to which strategy the agents adopted. In the post hoc analysis, cooperation towards the individualistic agents was lower than towards cooperative and tit-for-tat agents. Additionally, we analyzed the cooperation rate of the agents and we found the individualistic strategy led the agents to cooperate less compared to the other to Strategies, which suggests people might have reciprocated the autonomous agents to a certain extent Fig. 2. In our experiment, we could not find evidence that transparency affects people's behaviour.

Regarding **H2**, we have hypothesized that trust and group identification would be positively affected by transparent behaviour. On both measures, we found a significant interaction effect of transparency and strategy, which reveals the effect of transparency on trust and group identification was different across the three Strategies. In terms of the trust, the post-hoc analysis did not reveal a significant effect of transparency in any of the Strategies. However, the trends that are visible in Fig. 3 suggest this effect was negative for the cooperative agents and was positive for both the individualistic and tit-for-tat agents. In the post-hoc analysis for the group identification, we found a significant positive effect of transparency for the individualistic agents. For the remaining Strategies, similar trends are visible in Fig. 4 suggesting a negative effect for cooperative agents and a positive effect for tit-for-tat agents. Our hypothesis was only partially validated due to the fact that both group measures showed a positive effect only for two Strategies, the individualistic and tit-for-tat. Later in this section, we discuss the negative effect on the cooperative strategy.

In **H3**, we have predicted that transparent behaviours would positively affect the likeability and human-likeness of the agents. We only found a significant interaction effect between transparency and strategy on the perceived human-likeness. In other words, the effect of transparency on the perception of human-likeness was different across the three Strategies. Although the post hoc analysis did not reveal a significant effect of transparency in any of the Strategies, the trends suggest a negative effect on the cooperative agents, a positive effect on the individualistic agents and no effect is suggested for the tit-for-tat agents. In terms of likeability, we found a significant main effect of the strategy with the individualistic agents being significantly rated as less likeable compared to the cooperative and tit-for-tat agents Fig. 5. This hypothesis was validated in terms of human-likeness for the agents that use a individualistic strategy.

Our results suggest that adding transparent behaviour to an unconditional cooperator negatively affects the perceptions people have in terms of trust, group identification and human likeness. Although these differences were not statistically significant, the trends are congruent in the same direction. Further investigation is needed to support this claim. In terms of human-likeness, our intuition is that the unconditional cooperator might have revealed to the participants a non-optimal strategy, which a human would probably not do. However, the result for the group measures are counter-intuitive because the non-optimally of this strategy is related to the individual gains and it is not clear why the unconditional cooperator negatively affected then perception of the group.

7 Conclusions

Research in the field of artificial intelligence requires the design of system transparency able to improve the collaboration in human-agents and human-robot scenarios. This research discusses how strategy and transparency of artificial agents can influence human behavior in teamwork. Within the limits of the results found, we can state that transparency has significant effects on the trust, group identification and human likeness. This aspect turns out to be interesting in the context of public goods games and the design of relational and social capabilities in intelligent systems. Further research should consider the use of the Social Value Orientation [20] to randomize the sample between the condition before running the study. In addition, other type of transparency exploitation should be explored, as well as other game scenario and a more selected sample based on specific objectives, such as education or ecological sustainability. To conclude, a more comprehensive investigation of the methods to evaluate and implement the system transparency considering the effect of agents' strategy should be considered and tested in the wild.

References

1. Allen, K., Bergin, R.: Exploring trust, group satisfaction, and performance in geographically dispersed and co-located university technology commercialization teams. In: Proceedings of the NCIIA 8th Annual Meeting: Education that Works, pp. 18–20 (2004)
2. Axelrod, R.: On six advances in cooperation theory. Anal. Kritik **22**, 130–151 (2000). https://doi.org/10.1515/auk-2000-0107
3. Bartneck, C., Kulic, D., Croft, E., Zoghbi, S.: Measurement instruments for the anthropomorphism, animacy, likeability, perceived intelligence, and perceived safety of robots. Int. J. Soc. Robot. **1**, 71–81 (2008). https://doi.org/10.1007/s12369-008-0001-3
4. Bornstein, G., Nagel, R., Gneezy, U., Nagel, R.: The effect of intergroup competition on group coordination: an experimental study. Games Econ. Behav. **41**, 1–25 (2002). https://doi.org/10.2139/ssrn.189434

5. Burton-Chellew, M.N., Mouden, C.E., West, S.A.: Conditional cooperation and confusion in public-goods experiments. Proc. Nat. Acad. Sci. U.S.A. **113**(5), 1291–6 (2016)
6. Chen, J.Y.C., Lakhmani, S.G., Stowers, K., Selkowitz, A.R., Wright, J.L., Barnes, M.: Situation awareness-based agent transparency and human-autonomy teaming effectiveness. Theor. Issues Ergon. Sci. **19**(3), 259–282 (2018). https://doi.org/10.1080/1463922X.2017.1315750
7. Chen, J.Y., Barnes, M.J.: Human-agent teaming for multirobot control: a review of human factors issues. IEEE Trans. Hum.-Mach. Syst. **44**(1), 13–29 (2014)
8. Chen, J.Y., Barnes, M.J.: Agent transparency for human-agent teaming effectiveness. In: 2015 IEEE International Conference on Systems, Man, and Cybernetics, pp. 1381 1385. IEEE (2015)
9. Correia, F., et al.: Exploring prosociality in human-robot teams. In: 2019 14th ACM/IEEE International Conference on Human-Robot Interaction (HRI), pp. 143–151. IEEE (2019)
10. DARPA: Explainable artificial intelligence (XAI) program (2016). www.darpa.mil/program/explainable-artificial-intelligence,fullsolicitationatwww.darpa.mil/attachments/DARPA-BAA-16-53.pdf
11. Davis, D., Korenok, O., Reilly, R.: Cooperation without coordination: signaling, types and tacit collusion in laboratory oligopolies. Exp. Econ. **13**(1), 45–65 (2010)
12. Doshi-Velez, F., Kim, B.: Towards a rigorous science of interpretable machine learning (2017)
13. Fiala, L., Suetens, S.: Transparency and cooperation in repeated dilemma games: a meta study. Exp. Econ. **20**(4), 755–771 (2017)
14. Fredrickson, J.E.: Prosocial behavior and teamwork in online computer games (2013)
15. Fudenberg, D., Maskin, E.: The Folk theorem in repeated games with discounting or with incomplete. Information (2009). https://doi.org/10.1142/9789812818478_0011
16. Helldin, T.: Transparency for future semi-automated systems, Ph.D. dissertation. Orebro University (2014)
17. Herlocker, J.L., Konstan, J.A., Riedl, J.: Explaining collaborative filtering recommendations. In: Proceedings of the 2000 ACM Conference on Computer Supported Cooperative Work, pp. 241–250. ACM (2000)
18. Sedano, C.I., Carvalho, M., Secco, N., Longstreet, C.: Collaborative and cooperative games: facts and assumptions, pp. 370–376, May 2013. https://doi.org/10.1109/CTS.2013.6567257
19. Klien, G., Woods, D.D., Bradshaw, J.M., Hoffman, R.R., Feltovich, P.J.: Ten challenges for making automation a "team player" in joint human-agent activity. IEEE Intell. Syst. **19**(6), 91–95 (2004). https://doi.org/10.1109/MIS.2004.74
20. Lange, P., Otten, W., De Bruin, E.M.N., Joireman, J.: Development of prosocial, individualistic, and competitive orientations: theory and preliminary evidence. J. Pers. Soc. Psychol. **73**, 733–46 (1997). https://doi.org/10.1037//0022-3514.73.4.733
21. Leach, C.W., et al.: Group-level self-definition and self-investment: a hierarchical (multicomponent) model of in-group identification. J. Pers. Soc. Psychol. **95**(1), 144 (2008)
22. Lee, C.C., Chang, J.W.: Does trust promote more teamwork? Modeling online game players' teamwork using team experience as a moderator. Cyberpsychology Behav. Soc. Netw. **16**(11), 813–819 (2013). https://doi.org/10.1089/cyber.2012.0461. pMID: 23848999

23. McEwan, G., Gutwin, C., Mandryk, R.L., Nacke, L.: "i'm just here to play games": social dynamics and sociality in an online game site. In: Proceedings of the ACM 2012 Conference on Computer Supported Cooperative Work, CSCW 2012, pp. 549–558. ACM, New York (2012). https://doi.org/10.1145/2145204.2145289
24. Mercado, J.E., Rupp, M.A., Chen, J.Y., Barnes, M.J., Barber, D., Procci, K.: Intelligent agent transparency in human-agent teaming for multi-uxv management. Hum. Factors 58(3), 401–415 (2016)
25. Poursabzi-Sangdeh, F., Goldstein, D.G., Hofman, J.M., Vaughan, J.W., Wallach, H.M.: Manipulating and measuring model interpretability. CoRR arXiv:abs/1802.07810 (2018)
26. Rader, E., Cotter, K., Cho, J.: Explanations as mechanisms for supporting algorithmic transparency, pp. 103:1–103:13 (2018). https://doi.org/10.1145/3173574.3173677
27. Segal, U., Sobel, J.: Tit for tat: foundations of preferences for reciprocity in strategic settings. J. Econ. Theory 136(1), 197–216 (2007). https://EconPapers.repec.org/RePEc:eee:jetheo:v:136:y:2007:i:1:p:197-216
28. Zelmer, J.: Linear public goods experiments: a meta-analysis. Quantitative studies in economics and population research reports 361. McMaster University, June 2001. https://ideas.repec.org/p/mcm/qseprr/361.html

Explainable Robots

Explainable Multi-Agent Systems
Through Blockchain Technology

Davide Calvaresi[1]([⊠])(ID), Yazan Mualla[2], Amro Najjar[3], Stéphane Galland[2],
and Michael Schumacher[1]

[1] University of Applied Sciences and Arts Western Switzerland,
Sierre, Switzerland
{davide.calvaresi,michael.schumacher}@hevs.ch
[2] CIAD, Univ. Bourgogne Franche-Comté, UTBM, 90010 Belfort, France
{yazan.mualla,stephane.galland}@utbm.fr
[3] UMEA University, Umeå, Sweden
najjar@cs.umu.se

Abstract. Advances in Artificial Intelligence (AI) are contributing to
a broad set of domains. In particular, Multi-Agent Systems (MAS) are
increasingly approaching critical areas such as medicine, autonomous
vehicles, criminal justice, and financial markets. Such a trend is producing a growing AI-Human society entanglement. Thus, several concerns
are raised around user acceptance of AI agents. Trust issues, mainly due
to their lack of explainability, are the most relevant. In recent decades,
the priority has been pursuing the optimal performance at the expenses
of the interpretability. It led to remarkable achievements in fields such as
computer vision, natural language processing, and decision-making systems. However, the crucial questions driven by the social reluctance to
accept AI-based decisions may lead to entirely new dynamics and technologies fostering explainability, authenticity, and user-centricity. This
paper proposes a joint approach employing both blockchain technology
(BCT) and explainability in the decision-making process of MAS. By
doing so, current opaque decision-making processes can be made more
transparent and secure and thereby trustworthy from the human user
standpoint. Moreover, several case studies involving Unmanned Aerial
Vehicles (UAV) are discussed. Finally, the paper discusses roles, balance, and trade-offs between explainability and BCT in trust-dependent
systems.

Keywords: MAS · Goal-based XAI · Explainability · UAV ·
Blockchain

1 Introduction

Human decisions are increasingly relying on Artificial Intelligence (AI) techniques implementing autonomous decision making and distributed problem solving. Human-system interaction is pervading many domains, including health-care

© Springer Nature Switzerland AG 2019
D. Calvaresi et al. (Eds.): EXTRAAMAS 2019, LNAI 11763, pp. 41–58, 2019.
https://doi.org/10.1007/978-3-030-30391-4_3

[7], Cyber-Physical Systems [12,31], financial markets [38], and cloud computing [35]. Such entanglements enforced the ratification of the recent European General Data Protection Regulation (GDPR) law which underlines the right to explanations [14] and ACM US Public Policy Council (USACM)'s algorithmic transparency and accountability [1].

Therefore, the design of transparent and intelligible technologies is an impelling necessity. However, the interaction between autonomous AI-based systems (e.g., robots and agents) and humans decision processes raises concerns about the trust, reliability, and acceptance of autonomous systems. Recent studies proved that for both humans and software agents/robots, the trust into autonomous intelligent systems is strengthened if rules, decisions, and results can be explained. Hence, in the last decade, the hype about eXplainable Artificial Intelligence (XAI) [4,25] picked again. However, the majority of the recent studies focus on the interpretability and explanations for data-driven algorithms [5,18,24,41], thus still leaving open investigations concerning explainable agents and robots [4].

Humans tend to associate rationales to understanding and actions, developing a "mental states" [26]. A missing explanation can generate understanding that does not necessarily reflect AI's internal stance (self-deception). To a certain extent, dangerous situations may arise, putting the user safety at risk. According to the recent literature [5,37], explanations help users to increase confidence and trust, whereas misunderstanding the intentions of the intelligent system creates discomfort and confusion. Therefore, endowing these agents and robots with explainable behavior is paramount for their success. Interacting with these systems, however, there are domains and scenarios in which giving a proper explanation is **not** *(i)* possible, *(ii)* worth it, or *(iii)* enough. Therefore, the novelty proposed by this work is the following.

Contribution

This paper proposes to combine XAI, with blockchain technologies to ensure trust in domains where, due to environmental constraints or to some characteristics of the users/agents in the system, the effectiveness of the explanation may drop dramatically.

The rest of this article is organized as follows. Section 2 presents the background of this work in the domains of trust, explainability, and blockchain technology. Section 3 identifies three key research domains in which the synergy between BCT and XAI is necessary. Section 4 highlights the major challenges, Sect. 5 presents the proposed solution. Section 6 presents a use-case scenario, Sect. 7 discusses the scope of attainable solutions in which a combination of BCT and XAI is to be successful, and finally Sect. 8 concludes the paper.

2   Background

This section gives an overview of *trust* (Sect. 2.1), *explainability* (Sect. 2.2), and *blockchain* (Sect. 2.3) which are the key elements enabling the understanding of what their combination can provide to Multi-Agent Systems (MAS).

2.1 Trust

Autonomy is considered a basic feature for intelligent agents. Although it is highly desirable, such a property raises several challenges [40]. For example, *(i)* the agent designer must take into account the *autonomy of other agents* (run-time adaptation is a must for any agent to be competitive), and *(ii)* it is unrealistic to assume that other agents adopt a same/similar conduct.

Thus, artificial societies need some sort of control mechanisms. Traditionally, computational security has been claimed to be able to address a set of well-defined threats/attacks by relying on cryptography algorithms [21]. Yet, this approach requires the existence of a Trusted Third Party (TTP) to provide public and private keys and other credentials, which, for decentralized and open application scenario, becomes unrealistic [8]. On turn, several other *soft control* techniques have been defined to provide a certain degree of control without restricting the system development. These approaches rely on *social control mechanisms* (e.g., trust and reputation) that do not prevent undesirable events but ensure some social order in the system [15]. Nevertheless, they can allow the system to evolve in a way which prevents them from appearing again.

Several definitions have been proposed to define the notion of trust. Yet, the definition proposed by Gambetta *et al.* [23] is particularly useful and adopted by the MAS community.

"Trust is the subjective probability by which an agent A expects that another agent B performs a given action on which its welfare depends".

Therefore, trust is seen as an estimation or a prediction of the future or an expectation of an uncertain behavior, mostly based on previous behaviors [9]. A second form of trust is the *act of taking a decision* itself (e.g., relying on, counting on, or depending on the trustee). Summarizing, trust is both:

(i) a mental state about the other's trustworthiness (an evaluation) and
(ii) a decision or intention based on that evaluation [40]. To evaluate the trust, an agent relies on the *image* of the other agents. An image is an evaluative belief that tells whether the target is good or bad with respect to the given behavior. Images are results of internal reasoning from different sources of information that lead the agent to create a belief about the behavior of other agents [40].

2.2 Explainability

Explaining the decisions taken by an "intelligent system" has received relevant contributions from the AI community [16,27]. Earlier works on sought to build explainable expert systems. For this reason, after a prosperous phase, explainability received less attention in the 2000's. Recently, as AI systems are getting increasingly complex, explainable AI (XAI) reemerged to push for interpreting the "black-box" machine learning mechanisms and understanding the decisions of robots and agents. Consequently, research on XAI can be classified in two main branches:

- **Data-driven** (so-called *perceptual* [36]) XAI:
  It aims at *interpreting* the results of "black-box" machine learning mechanisms such as Deep Neural Networks (DNN) [48]. This research achieved intriguing results (e.g., understanding why a DNN mistakenly labelled a tomato as a dog [44]). Therefore, the lust to *interpret*, or *provide a meaning* for an obscure machine learning model (whose inner-workings are otherwise unknown or non-understandable by the human observer) is tickling the researchers.
- **Goal-driven** (so-called *cognitive* [36]) XAI:
  Research from cognitive science has shown that humans attribute mental states to robots and autonomous agents. This means that humans tend to attribute goals, intentions and desires to these systems. This branch of XAI aims at explaining the rationales of the decisions of intelligent agents and robots by citing their goals, beliefs, emotions, etc. [4]. Providing such explanations allows the human to understand *capabilities*, *limits*, and *risks* of the agent/robot they are interacting with, and thereby raising the user awareness and trust in the agent, facilitating *critical decisions* [4,13].

2.3   Blockchain Technology

Blockchain is a distributed technology employing cryptographic primitives that rely on a *(i)* membership mechanism, and *(ii)* a consensus protocol to maintain a shared, immutable, and transparent append-only register [9]. Observing The information (digitally signed transactions) delivered by the entities part of the network are grouped into blocks chronologically time-stamped.

The single block is identified by a unique block-identifier, which is obtained by applying a hash function to its content and it is stored in the subsequent block. Such a technique is part of a set of mechanisms considered *tamper-proof* [8] modification of the content of a block, can be easily verified by hashing it again, and comparing the results with the identifier from the subsequent block. Moreover, depending on the distribution and consensus mechanism, the blockchain can be replicated and maintained by every (or a sub-set) participant(s) (so-called peers). Thus, a malicious attempt to tamper the information stored in the registry can be immediately spotted by the participants, thus guaranteeing immutability of the ledger [8]. Several technological implementations of the blockchain can execute arbitrary tasks (so-called smart contracts) allowing the implementation of desired functionality. Alongside the blocks, such smart contracts represent the logic applied and distributed with the data [28].

Technology. BCT can be distinguish between *permissionless* and *permissioned* (public and private) blockchain systems [43]:

- A blockchain is *permissionless* when the identities of participants are either pseudonymous or anonymous (every user can participate in the consensus protocol, and therefore append a new block to the ledger).

– A blockchain is permissioned if the identities of the users and rights to participate in the consensus (writing to the ledger and/or validating the transactions) are controlled by a membership service.

Moreover, on the one hand, a **permissioned** blockchain is *public* when any-
one can read the ledger, **but** only predefined set of users can participate in the
consensus. On the other hand, it is *private* when even the right to read the ledger
is controlled by a membership/identity service.

3 Application Domains

Trust is still an outstanding open challenge in the area of intelligent systems.
However, **Blockchain** technology and techniques derived from the **XAI** discipline can be tightly coupled to provide reconciling, feasible, and cost-effective
solutions. On the one hand, explainable behaviors can enable the trustor to
evaluate the soundness and the completeness of the actions of the trustee, and
thereby it can evaluate its competences, and examine the rationale behind its
behavior. On the other hand, BCT can allow the trustor to unequivocally assess
the *reputation* of the trustee based on existing history knowledge about it. In
this paper, we explore reconciling solutions combining both XAI and BCT. This
synergy can be beneficial for several application domains involving collaborations
among agents to undertake joint decisions in a decentralized manner. Below, we
identify three types of applications in which such a synergy would be highly
beneficial.

Cloud Computing is a distributed ecosystem involving multiple actors each concerned with accomplishing a different set of goals. Agent-based systems have
been underlined as a platform capable of adding intelligence to the cloud ecosystem and allowing to undertake critical tasks such as resource management in a
decentralized manner that considers the distributed and multi-partite nature of
the cloud ecosystem [45]. In a typical three partite scenario, it involves: *(i)* Cloud
providers who seek to offer an adequate Quality of Service (QoS) while minimizing the energy consumption and maintenance costs of its data-centers [22], *(ii)*
Cloud users whose aim is to minimize the cost they pay to the provider while fur-
nishing a satisfactory service to their end-users [34], and brokers. In exchange for
a fee, a broker reserves a large pool of instances from cloud providers and serves
users with price discounts. Thus, it optimally exploits both pricing benefits of
long-term instance reservations and multiplexing gains [47]. In such a scenario,
given the multitude of providers, brokers and offers available in the cloud market,
both explainability and trust are critical to help these actors make their strategic
decisions. For instance, when recommending resources from a particular cloud
provider, a broker could rely on BCT technology to assess the reputation and the
trustworthiness of the provider. Several important data could be inscribed on
the ledger including the availability, reliability and the average response time of
the virtual instances leased from this provider. When giving a recommendation,
the broker might also use explainability to provide a transparent service to its

client and explain why some specific decision were made (e.g., the choice of one provider) and why some un-expected events took place (e.g., an SLA violation).

Smart Cities. The densely populated smart cities are administrated by several governmental and civil society actors, where vivid economic services involving a multitude of individual stakeholders take place. In such services, the use of agents for Unmanned Aerial Vehicles (UAVs) is gaining more interest especially in complex application scenarios where coordination and cooperation are necessary [32]. In particular, in the near future, UAVs will require access to an inter-operable, affordable, responsive, and sustainable networked system capable of providing service, joint, inter-agency, and real-time information exchanges. Such systems must be distributed, scalable, and secure. The main components are human interfaces, software applications, network services, information services, and the hardware and interfaces necessary to form a complete system that delivers secured UAVs operations [28]. Recalling that BCT allows creating a peer-to-peer decentralized network with an information protection mechanism [3], such a network can provide secure communication system within the MAS [20], thus operating as distributed control and secure system to ensure the trust among UAVs and other actors.

User Satisfaction Management. Agents are autonomous entities bound to individual perspectives, for these reasons, user agents were used to represent user satisfaction [35]. However, end-user satisfaction is known to be subjective [33] and influenced by several Influence Factors (IF) [39], including Human IFs (e.g., expertise, age, personality traits, etc.), Context IF (e.g., expectations) and System IFs (i.e., the technical properties of the systems used to consume the service). Both XAI and BCT can have key contributions helping agents overcome these challenges and improve user satisfaction. On the one hand, explainability enables the agent to provide convincing recommendations to the user by showing that the agent's decisions were in line with the user preferences. On the other hand, BCT can play an important role in assuring both the user and her agent that privacy and authentication measures are integrated to protect the user preferences and private data from exploitation.

4   Challenges

The combination of MAS, BCT, and XAI can be particularly strategic in several application fields. Real-world scenarios are often characterized by a combination of limited resources such as computational capability, memory, space, and in particular *time* [11,12,19]. Therefore, Sect. 4.1 tackles the application of the proposed solution in Resource-Constrained (RC). Another relevant dimension characterizing real-world application is the *trust* in the systems or in their components [9,10,40]. Thus, Sect. 4.2 addresses the Lack of Trust (LT) as main driver.

4.1 RC Scenarios

In real-world applications, systems must cope with a bounded availability of resources. On the one hand, we can mention tangible resources such as memory, computational capability, and communication bandwidth [12]. On the other hand, we can have reputation, trust, and time. The latter is crucial especially in safety-critical scenarios, when failing to deliver a result in/on time might have catastrophic consequences [6].

A possible example can be a UAVs firefighting scenario.

Let us assume that a UAV detects a fire in a nearby woods, and that the fire has already spread to an extent unmanageable by a single UAV. The only viable option for the UAV which detected the fire is to ask for support from the firefighting center, managed by humans, to send other UAVs. This requires the UAV to explain the situation to the representative human in the firefighting center. Considering that such a situation needs an intervention as prompt as possible, the UAV requesting assistance cannot *produce* and *deliver* an "extensive" explanation for its requests, plans, and the consequences of possible inaction. Achieving a consensus on an over-detailed (for the situation) explanation would be unaffordably time-consuming, thus leading to potentially considerable losses. A possible solution is to enable the requester to rely on BCT, which can ensure its possible trustworthiness (e.g., via reputation) and authenticity, compensating a less detailed explanation leading to a faster reaction to handle the fire.

4.2 LT Scenarios

In scenarios where time is not critical, the opportunity is given to an agent with low reputation to express itself to increase the trust with other actors. For example, a swarm of UAVs can be created to perform tasks that cannot be performed by one UAV or to increase the efficiency of a specific task. In such situations, there is a need for a mechanism for UAVs to join a swarm. Yet, a UAV with a low reputation may find it difficult to join a swarm. With explainability, it is possible that swarm management gives this UAV a chance to express itself in order to increase its trust and hence its chances to be accepted in the swarm. Another example is when it is not possible to determine the reputation of a UAV due to the inability to access the blockchain. This UAV can be given the chance of explaining its goals to increase the likelihood of an agreement.

5 Proposed Solution

According to the application domains and scenarios presented in Sect. 4, a two-folded solution (for RC and LT scenarios) follows.

5.1 RC

In scenarios in which the operating *time* is constrained (Sect. 4.1) and delivering a complete and high-quality explanation is not viable, the quality and granularity of a given explanation might be degraded to still comply with the timing constraints.

Similarly, if the *understanding capability* of the recipient of a given explanation is limited (Scenarios 3 and 4 in Table 1), the quality of the explanation can be lowered (since it might not be understood/appreciated) saving both time and effort (e.g., computational capability, memory).

Lower quality explanations are characterized by less details (coarse-grained) or unfaithful explanations. While offering brief insights on how and why a decision was taken, coarse-grained explanations do not provide a fully detailed explanation unless this is explicitly demanded by the explainee. Unfaithful explanation do not respect the actual mechanism that led to a given decision. Instead, their aim is to provide an understandable and easy explanation. A possible way of providing unfaithful explanation is relying on contrastive explanations. The latter consist of justifying one action by explaining why alternative actions were not chosen. While contrastive explanations do not necessarily describe the decision-making process of the agent, recent research has shown that they can be easily produced and easily understandable by the user [29]. Therefore, both coarse-grained and unfaithful explanations convey the message, thus accomplishing the explicative intent. Since an effective explanation might not be the most precise or faithful, it is possible to infer that precision and effectiveness of an explanation can be decorrelated. On the one hand, if the principal objective is to share the rationale behind a given decision, opting for an effective and potentially less precise explanation might be the best option [4]. On the other hand, if transparency is a mandatory requirement, a detailed and faithful explanation must be provided. For example, *time* available to produce and provide an explanation in a given context/situation is a factor influencing the agent, thus possibly impacting on the faithfulness of its explanations. In case the amount of time is too constrictive, the agent might opt for a short, simple, and unfaithful explanation (even though a detailed one would be preferred). Moreover, depending on time available, context, and explainee, the explainer may attempt at explaining the same concept employing different types of data or same data but with different granularity and complexity (e.g., images, text, or raw data).

To lower the explanation quality/granularity, without affecting the trust (information-, user-, or agent-wise), we propose to enforce the provided explanation with BCT. By doing so, we would compensate a less effective explanation with the guarantees provided by BCT technology, still keeping the system running and the trust unaffected by a time-critical scenario.

Table 1 lists four possible situations we have identified. Beside the *Time Available*, expressed in seconds, the other features are represented by adimensional numbers (useful to provide a quick and synthetic overview). *Ratio*, stands for correlation between the *quality* of a given explanation (possibly combined with the support of BCT) and how it is *understood, perceived* or *accepted* (if relying more on the BCT then on the actual explanation) by the recipient.

Table 1. Possible combinations of explanations' quality and blockchain support with the recipient's capabilities of understanding.

Scenario	Time available (seconds)	Explanation quality	Recipient understanding	Blockchain support	Gain
1	10	10	10	0	10/10
2	5	7	10	3	10/10
3	10	10	5	0	10/5
4	10	2	5	3	5/5

Scenario 1 the first scenario reproduces an ideal situation: having *(i)* enough time to provide a comprehensive and solid explanation and *(ii)* a recipient who has time and can process/understand the provided explanation. In this case, the support of the BCT is not necessary.

Scenario 2 short in time, and with a recipient able to fully understand and process the explanation, the agent opts for degrading the quality of the explanation relying on the contribution of BCT. In this case, the recipient's decision might not be affected by the lack of granularity of the received explanation.

Scenario 3 although the available time is enough to produce a robust explanation, the recipient is not able to entirely understand/process it. Therefore, since the explanation goes already over its purpose, it is not necessary to employ the BCT.

Scenario 4 the available time might be more than enough to produce a robust explanation, which however goes beyond the understanding capability of the recipient. Therefore, to save time and resources, the explanation can be degraded and coupled with the support of BCT, enough to match the recipient expectation and capability.

5.2   LT

In circumstance where an agent/user has a reputation lower than a given threshold, it can be labelled as not fully trustworthy. In this condition, although the user/agent might be able to provide an excellent explanation, it could not be trusted, or it could not get a chance to express it. Therefore, binding the explanation with BCT might relieve the agent explaining from the burden of a low reputation (obtained for a whatever *unfortunate* reason in a precedent point in time). Such a solution/approach can be associated to the famous dilemma "The Boy Who Cried Wolf" narrated by Aesop [2], the well-known Greek fabulist and storyteller. The fable narrates of a young shepherd who, just for fun, used to fool the gentlemen of the nearby village making fake claims of having his flock attacked by wolves. However, when the wolves attacked the flock for real, the villagers did not respond to the boy's cries (since they considered it to be just another false alarm). Therefore, the wolves end up ravaging the entire flock. This story is used as an example of the serious consequences of spreading false claims and alarms, generating mistrust in the society and resulting in the subsequent disbelieving the true claims. To "cry wolf", a famous English idiom glossed in Oxford English Dictionary [17], was derived from this fable to reflect the meaning of *spreading false alarms.*

Such a moral, applied to *human* societies, can also be applied to *agent* societies. For example, the requests of a UAV with a low reputation might be neglected because its records on the the ledger showed that it has been issuing false alarms about fires in the woods. However, with the possibility of explaining its new alarms and supporting its claims with tangible proofs (*e.g.,* images and footage from the fire location), if its explanations were convincing enough, the UAV might be able to overcome (and improve) its low reputation.

The next section addresses the UAVs package delivery, which is a use case from the real world. In such a scenario, multiple UAVs need to coordinate in order to achieve a common goal. To do so, members of the same UAV team (i.e., swarm) should share a common understanding and maintain a trustworthy relationship. To address these concerns, potentially time-constrained, the following section studies UAVs interaction and reputation by employing explainability and BCT.

6 Explainability and BCT: The UAVs Package Delivery Use Case

In 02 Aug 2018, the U.S. Patent and Trademark Office issued a new patent for retail giant Walmart seeking to utilize BCT to perfect a smarter package delivery tracking system [42]. Walmart describes a "smart package" delivered by a UAV that includes a device to record information about a blockchain related to the content of the package, environmental conditions, location, manufacturer, model number, etc. The application states that the blockchain component will be encrypted into the device and will have "key addresses along the chain of the package's custody, including hashing with a seller private key address, a courier private key address and a buyer private key address" [46].

Typically, modeled as agents, UAVs can be organized in swarms to help them achieve more than what they could solely. A decentralized swarm management system can add or remove UAVs from the swarm. To join the swarm, a reputation threshold should be acquired by the UAV. In cases of low reputation UAV (Sect. 5.2), the choice is given to the UAV to explain the reasons it must join the swarm.

UAVs use voting in the swarm to decide decisions like adding/removing UAVs, tasks to perform, etc. Before each vote, the possibility is given to each UAV to explain what it considers the best for the swarm in terms of what goals to achieve and how to do them. The swarm management system has a blockchain distributed ledger that is connected to Internet through various wireless networks (e.g., WiFi, 4G/5G, satellite). It allows the swarm to check the reputation of any UAV willing to join the swarm as well as the reputation of any outer actors that wish to communicate with the swarm.

For example, suppose that a new UAV has joined the swarm and is granted a private key. Once the UAV exists on the blockchain distributed ledger of the swarm management system, the levels of access, control, and/or authority are determined for the new UAV.

External actors (UAVs or people) may ask the swarm to perform tasks for them. Negotiation will commence between the external actor and the swarm that considers the trade-off between explainability and reputation of the actor, the profit of performing the task (in case of commercial swarms), the general welfare (in case of governmental or non-profit organizations). If the swarm accepts to perform a given task, smart contracts can be used to transfer commands between agents in the form of data or executable code in real-time.

Let us assume that an actor (human, device, etc.) in a smart home asks the swarm to make a delivery order. Depending on the time window of the delivery transaction, different scenarios that combine reputation and explainability are considered (Sect. 5.1). Figure 1 shows the steps to consider as per the constraints of the scenario.

If an agreement is reached, a smart contract is generated with the order data (e.g., package characteristics, client data, location, and designated UAV) and the information is sent to the Blockchain. Then, the UAV commits a transaction to the traffic coordinator to provide an air corridor for it and a new smart contract is concluded between them.

The UAV starts the delivery to the smart home. Once near the smart home, the UAV will contact the smart window using a wireless network. The smart window is connected to the internet as any other device in the smart home. This allows it to ask the blockchain if it recognizes and verifies this UAV and its swarm, and if it is the swarm that singed the smart contract. If the UAV is trustworthy, the window will open to allow it to drop the package. When the delivery is completed, the UAV notifies the traffic coordinator that the air corridor is no longer needed.

To achieve all of that, there is a need for defining two important aspects. First, protocols for the registration, verification, peer-to-peer interaction of the UAVs. Second, smart contracts between the swarm and any other actor in the environment (UAV, device, human, etc.), that govern the services used or provided by the swarm. Moreover, the use of a blockchain infrastructure helps in identifying misbehaving UAVs by multiple parties and such activities are recorded in an immutable ledger. These misbehaving assessments may be performed by analytical algorithms or machine learning models performed off-chain and interfaced with the blockchain ledger through smart contracts. Once determined, the misbehaving UAV will be given the chance to explain its behaviour and actions in the after-action phase (Sect. 5.2).

Of course, the service provided by the UAV will affect the weights of importance for the reputation and explainability. For example, in time critical situations, there is no time for long/complex explanation, and the reputation plays the more significant role.

7   Discussion

Analyzing the solutions proposed in Sect. 5, Fig. 1 summarizes the possible outcomes eliciting the attainable solutions.

In particular, time availability is the predominant factor. If an agent is short in time, explainability might not be an option. Therefore, the agent is demanded to have a trustworthy reputation (proved by the BCT) to achieve a possible agreement. In the case no explanation can be provided and the reputation value is below an acceptable threshold there is no possible solution, and the request of the agent (as we saw in the UAV example above), is rejected.

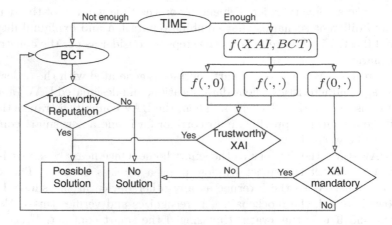

Fig. 1. Decision process wrt available time, explanation, and reputation.

If the available time to produce an explanation is enough, explainability becomes an option. The agent can rely on $f(XAI, BCT)$, a combination of explainability and BCT. The agent might rely only on explainability $f(\cdot, 0)$, only on BCT $f(0, \cdot)$, or on any given combination of both $f(\cdot, \cdot)$. In the latter case, the weights composing this combination mainly depend on the *(i)* specific context, *(ii)* nature of the problem to be explained, *(iii)* explanation capability of the agent and on *(iv)* understanding capability of the agent receiving the explanation. Moreover, on the one hand, having explainability might be necessary and enforced by law. On the other hand, low reputation/trustworthiness of an agent cannot be ignored even if it provided an adequate explanation.

Figure 2 shows a sequence of interaction within a society of agents, the first agent $A1$ attempts to send an explanation to agent $A2$. Depending on the scenario, $A1$ might possibly be short in time, might possibly be able to rely on BCT for reputation. Based on the explanation/reputation submitted by $A1$ to $A2$, the

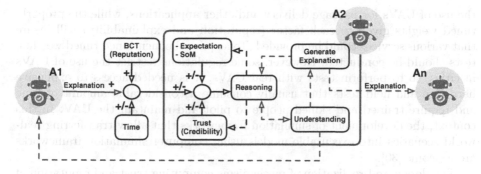

Fig. 2. Representation of the explanation life-cycle. (Color figure online)

latter would be able to assess the trustworthiness of $A1$, compare the behavior of $A1$ with its own expectation, and define/update a State of Mind (SoM) about $A1$ intentions. As a result of this reasoning process, $A2$ (delineated by the blue box) builds an understanding of $A1$ and its explanation. Such an understanding is then used to: *(i)* generate an explanation describing $A1$ behavior and communicate it with other agents A_n, *(ii)* refine $A2$'s SoM, reasoning, and expectations about $A1$, and *(iii)* possibly coming back to $A1$ to ask more details/clarifications about its explanation.

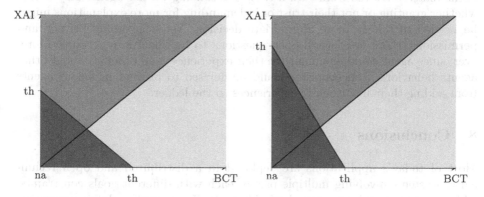

Fig. 3. Symmetric and Asymmetric XAI and BCT contributions (Color figure online)

Figure 3 illustrates the possible synergy between XAI and BCT. The blue diagonal line represents the threshold delineating whether the combination of XAI and BCT satisfies the minimal requirements (area in green) or not (area in red). In the left figure (symmetric case), the contributions from XAI and BCT are equal, thus symmetric. In the right figure (asymmetric case), a contribution from either XAI or BCT has higher impact (XAI in the example in the figure).

In the domain of UAVs, where the regulations are not mature enough [32], the combination of reputation and explainability will increase the trust of clients in the use of UAVs for package delivery and other applications, while the properly tuned weights given to each factor (reputation and explainability) will insure that various services could be provided. To acquire the mentioned tuned weights, tests should be conducted. However, some regulations restrict the use of UAVs in cities, so to perform tests with real UAVs, it is needed access to expensive hardware and field tests that usually consume a considerable amount of time and require trained and skilled people to pilot and maintain the UAV. In this context, the development of simulation frameworks that allow transferring real-world scenarios into executable models using computer simulation frameworks are welcome [30].

The design and realization of mechanisms computing trust and reputation of agents communities via blockchain are strictly dependent on the application scenarios and available technologies. Therefore, they are delegated to future studies. Nevertheless, at the current stage, it is possible to provide key research directions. For example, as mentioned in Sect. 2.1, to undertake the trust evaluation process, agents rely on the social *image* of other agents [40]. An agent constructs such an image relying on *(i) direct experiences* (e.g., direct interactions between the trustor and the trustee), *(ii) communicated experiences* (e.g., interactions between the trustee and another agent communicated to the trustor), and *(iii) social information* (e.g., monitoring social relations and position of the trustee in the society). The interactions and mechanism enabling the computation of trust and reputation can be stored on a blockchain. Thus, depending on the agent *image* retrieved from such a trusted technology, other agents may decide whether granting or not their trust or if demanding for more explanations might be needed to take a more appropriate decision. Yet, concerning privacy and permissions, there are several open questions to be taken into account. Moreover, since agent could communicate their experiences and opinions about other agents behaviors, a mechanism should be devised to prevent malicious agents from adding their unauthentic experiences to the ledger.

8 Conclusions

Most of today's applications are deployed in a distributed and open technical ecosystems involving multiple parties each with different goals constraints. This paper proposed an approach combining BCT and explainability supporting the decision-making process of MAS. Such an approach can remove the current opaqueness of decision-making processes making them interpretable and trustworthy from both agent and human user point of views. It is worth to recall that explainability allows collaborating parties to express their intentions and reach common understandings and that BCT offers a decentralized authentication mechanisms capable of ensuring trust and reputation management. Then, it identified some applications where the contribution of these technologies revealed to be crucial. Three scenarios have been identified: *(i)* BCT are strictly necessary, *(ii)* explainability is mandatory, and *(iii)* a combination of them is possible and subject to a threshold function. Moreover, several practical case study involving UAVs have been discussed, analyzing roles, balance, and trade-offs between explainability and BCT in trust-dependent systems. This work is an initial step towards building synergies between explainable AI and BCT. The future work is to *(i)* investigate a MAS model suitable for XAI and BCT, *(ii)* design and develop a MAS framework to implement explainable and BCT dynamics, *(iii)* realize smart contracts supporting an efficient communication among light weight devices, *(iv)* assess a possible interdependence among explainability and BCT (in particular involving remote robots such as UAV and HGV), and *(v)* apply and study the developed solutions to UAVs swarm.

Acknowledgments. This work is partially supported by the Regional Council of Bourgogne Franche-Comté (RBFC, France) within the project UrbanFly 20174-06234/06242. This work is partially supported by the Wallenberg AI, Autonomous Systems and Software Program (WASP) funded by the Knut and Alice Wallenberg Foundation.

References

1. ACM US: Public policy council: statement on algorithmic transparency and accountability (2017)
2. Aesop: Aesop's Fables. OUP, Oxford (2002)
3. Ali, M., Nelson, J.C., Shea, R., Freedman, M.J.: Blockstack: a global naming and storage system secured by blockchains. In: USENIX Annual Technical Conference, pp. 181–194 (2016)
4. Anjomshoae, S., Najjar, A., Calvaresi, D., Framling, K.: Explainable agents and robots: results from a systematic literature review. In: Proceedings of the 18th International Conference on Autonomous Agents and Multi-Agent Systems (AAMAS) (2019)
5. Biran, O., Cotton, C.: Explanation and justification in machine learning: a survey. In: IJCAI 2017 Workshop on Explainable AI (XAI), p. 8 (2017)
6. Buttazzo, G.C.: Hard Real-Time Computing Systems: Predictable Scheduling Algorithms and Applications, vol. 24, 3rd edn. Springer, Heidelberg (2011). https://doi.org/10.1007/978-1-4614-0676-1
7. Calvaresi, D., Cesarini, D., Sernani, P., Marinoni, M., Dragoni, A.F., Sturm, A.: Exploring the ambient assisted living domain: a systematic review. J. Ambient Intell. Humaniz. Comput. 8(2), 239–257 (2017)
8. Calvaresi, D., Dubovitskaya, A., Calbimonte, J.P., Taveter, K., Schumacher, M.: Multi-agent systems and blockchain: results from a systematic literature review. In: Demazeau, Y., An, B., Bajo, J., Fernández-Caballero, A. (eds.) PAAMS 2018. LNCS (LNAI), vol. 10978, pp. 110–126. Springer, Cham (2018). https://doi.org/10.1007/978-3-319-94580-4_9
9. Calvaresi, D., Dubovitskaya, A., Retaggi, D., Dragoni, A., Schumacher, M.: Trusted registration, negotiation, and service evaluation in multi-agent systems throughout the blockchain technology. In: International Conference on Web Intelligence (2018)
10. Calvaresi, D., Leis, M., Dubovitskaya, A., Schegg, R., Schumacher, M.: Trust in tourism via blockchain technology: results from a systematic review. In: Pesonen, J., Neidhardt, J. (eds.) Information and Communication Technologies in Tourism 2019, pp. 304–317. Springer, Cham (2019). https://doi.org/10.1007/978-3-030-05940-8_24

11. Calvaresi, D., Marinoni, M., Dragoni, A.F., Hilfiker, R., Schumacher, M.: Real-time multi-agent systems for telerehabilitation scenarios. Artif. Intell. Med. **96**, 217–231 (2019)
12. Calvaresi, D., Marinoni, M., Sturm, A., Schumacher, M., Buttazzo, G.C.: The challenge of real-time multi-agent systems for enabling IoT and CPS. In: Proceedings of the International Conference on Web Intelligence, Leipzig, Germany, 23–26 August 2017, pp. 356–364 (2017). https://doi.org/10.1145/3106426.3106518
13. Calvaresi, D., Mattioli, V., Dubovitskaya, A., Dragoni, A.F., Schumacher, M.: Reputation management in multi-agent systems using permissioned blockchain technology (2018)
14. Carey, P.: Data Protection: A Practical Guide to UK and EU law. Oxford University Press Inc., Oxford (2018)
15. Castelfranchi, C.: Engineering social order. In: Omicini, A., Tolksdorf, R., Zambonelli, F. (eds.) ESAW 2000. LNCS (LNAI), vol. 1972, pp. 1–18. Springer, Heidelberg (2000). https://doi.org/10.1007/3-540-44539-0_1
16. Chandrasekaran, B., Tanner, M.C., Josephson, J.R.: Explaining control strategies in problem solving. IEEE Intell. Syst. **1**, 9–15 (1989)
17. Oxford English Dictionary: Compact Oxford English Dictionary. Oxford University Press, Oxford (1991)
18. Doran, D., Schulz, S., Besold, T.R.: What does explainable AI really mean? A new conceptualization of perspectives. arXiv preprint arXiv:1710.00794 (2017)
19. Dragoni, A.F., Sernani, P., Calvaresi, D.: When rationality entered time and became a real agent in a cyber-society. In: Proceedings of the 3rd International Conference on Recent Trends and Applications in Computer Science and Information Technology, RTA-CSIT 2018, Tirana, Albania, 23rd–24th November 2018, pp. 167–171 (2018). http://ceur-ws.org/Vol-2280/paper-24.pdf
20. Castelló Ferrer, E.: The blockchain: a new framework for robotic swarm systems. In: Arai, K., Bhatia, R., Kapoor, S. (eds.) FTC 2018. AISC, vol. 881, pp. 1037–1058. Springer, Cham (2019). https://doi.org/10.1007/978-3-030-02683-7_77
21. Forouzan, B.A.: Cryptography & Network Security, 1st edn. McGraw-Hill Inc., New York, NY, USA (2008)
22. Fox, A., et al.: Above the clouds: a Berkeley view of cloud computing. Department of Electrical Engineering and Computer Sciences, University of California, Berkeley, Report UCB/EECS 28(13), 2009 (2009)
23. Gambetta, D., et al.: Can we trust trust. Trust: Making and breaking cooperative relations, vol. 13, pp. 213–237 (2000)
24. Guidotti, R., Monreale, A., Ruggieri, S., Turini, F., Giannotti, F., Pedreschi, D.: A survey of methods for explaining black box models. ACM Comput. Surv. (CSUR) **51**(5), 93 (2018)
25. Gunning, D.: Explainable artificial intelligence (XAI). Defense Advanced Research Projects Agency (DARPA), ND Web (2017)
26. Hellström, T., Bensch, S.: Understandable robots-what, why, and how. Paladyn J. Behav. Robot. **9**(1), 110–123 (2018)
27. Kass, R., Finin, T., et al.: The need for user models in generating expert system explanations. Int. J. Expert Syst. **1**(4), 345–375 (1988)
28. Kuzmin, A., Znak, E.: Blockchain-base structures for a secure and operate network of semi-autonomous unmanned aerial vehicles. In: 2018 IEEE International Conference on Service Operations and Logistics, and Informatics (SOLI), pp. 32–37. IEEE (2018)
29. Miller, T.: Explanation in artificial intelligence: insights from the social sciences. Artif. Intell. **267**, 1–38 (2018)

30. Mualla, Y., Bai, W., Galland, S., Nicolle, C.: Comparison of agent-based simulation frameworks for unmanned aerial transportation applications. Proc. Comput. Sci. **130**(C), 791–796 (2018)

31. Mualla, Y., Najjar, A., Boissier, O., Galland, S., Tchappi, I., Vanet, R.: A cyber-physical system for semi-autonomous oil&gas drilling operations. In: 5th Workshop on Collaboration of Humans, Agents, Robots, Machines and Sensors. Third IEEE International Conference on Robotic Computing (2019)

32. Mualla, Y., et al.: Between the megalopolis and the deep blue sky: challenges of transport with uavs in future smart cities. In: Proceedings of the 18th International Conference on Autonomous Agents and MultiAgent Systems. International Foundation for Autonomous Agents and Multiagent Systems (2019)

33. Najjar, A.: Multi-agent negotiation for QoE-aware cloud elasticity management. Ph.D. thesis, École nationale supérieure des mines de Saint-Étienne (2015)

34. Najjar, A., Serpaggi, X., Gravier, C., Boissier, O.: Survey of elasticity management solutions in cloud computing. In: Mahmood, Z. (ed.) Continued Rise of the Cloud. CCN, pp. 235–263. Springer, London (2014). https://doi.org/10.1007/978-1-4471-6452-4_10

35. Najjar, A., Serpaggi, X., Gravier, C., Boissier, O.: Multi-agent systems for personalized QoE-management. In: 2016 28th International Teletraffic Congress (ITC 28), vol. 3, pp. 1–6. IEEE (2016)

36. Neerincx, M.A., van der Waa, J., Kaptein, F., van Diggelen, J.: Using perceptual and cognitive explanations for enhanced human-agent team performance. In: Harris, D. (ed.) EPCE 2018. LNCS (LNAI), vol. 10906, pp. 204–214. Springer, Cham (2018). https://doi.org/10.1007/978-3-319-91122-9_18

37. Nomura, T., Kawakami, K.: Relationships between robot's self-disclosures and human's anxiety toward robots. In: Proceedings of the 2011 IEEE/WIC/ACM International Conferences on Web Intelligence and Intelligent Agent Technology-Volume 03, pp. 66–69. IEEE Computer Society (2011)

38. Parkes, D.C., Wellman, M.P.: Economic reasoning and artificial intelligence. Science **349**(6245), 267–272 (2015)

39. Reiter, U., et al.: Factors influencing quality of experience. In: Möller, S., Raake, A. (eds.) Quality of Experience. TSTS, pp. 55–72. Springer, Cham (2014). https://doi.org/10.1007/978-3-319-02681-7_4

40. Sabater-Mir, J., Vercouter, L.: Trust and reputation in multiagent systems. In: Multiagent Systems, p. 381 (2013)

41. Samek, W., Wiegand, T., Müller, K.R.: Explainable artificial intelligence: understanding, visualizing and interpreting deep learning models. arXiv preprint arXiv:1708.08296 (2017)

42. Simon, J., et al.: Managing participation in a monitored system using blockchain technology. US Patent Application 15/881,715, 2 August 2018

43. Swanson, T.: Consensus-as-a-service: a brief report on the emergence of permissioned, distributed ledger systems (2015)

44. Szegedy, C., et al.: Intriguing properties of neural networks. arXiv preprint arXiv:1312.6199 (2013)

45. Talia, D.: Clouds meet agents: toward intelligent cloud services. IEEE Internet Comput. **16**(2), 78–81 (2012)

46. Walmart Retail Company: Walmart wants blockchain to make shipping 'smarter'. https://mrtech.com/news/walmart-wants-blockchain-to-make-shipping-smarter/. Accessed March 2018

47. Wang, W., Niu, D., Li, B., Liang, B.: Dynamic cloud resource reservation via cloud brokerage. In: 2013 IEEE 33rd International Conference on Distributed Computing Systems, pp. 400–409. IEEE (2013)
48. Zhang, Q., Zhu, S.C.: Visual interpretability for deep learning: a survey. Front. Inf. Technol. Electron. Eng. **19**(1), 27–39 (2018)

Explaining Sympathetic Actions
of Rational Agents

Timotheus Kampik(✉) ⓘ, Juan Carlos Nieves ⓘ, and Helena Lindgren ⓘ

Umeå University, 901 87 Umeå, Sweden
{tkampik,jcnieves,helena}@cs.umu.se

Abstract. Typically, humans do not act purely *rationally* in the sense
of classic economic theory. Different patterns of human actions have been
identified that are not aligned with the traditional view of human actors
as rational agents that act to maximize their own utility function. For
instance, humans often act sympathetically – i.e., they choose actions
that serve others in disregard of their egoistic preferences. Even if there
is no immediate benefit resulting from a sympathetic action, it can be
beneficial for the executing individual in the long run. This paper builds
upon the premise that it can be beneficial to design autonomous agents
that employ sympathetic actions in a similar manner as humans do. We
create a taxonomy of sympathetic actions, that reflects different goal
types an agent can have to act sympathetically. To ensure that the sympathetic actions are recognized as such, we propose different explanation
approaches autonomous agents may use. In this context, we focus on
human-agent interaction scenarios. As a first step towards an empirical
evaluation, we conduct a preliminary human-robot interaction study that
investigates the effect of explanations of (somewhat) sympathetic robot
actions on the human participants of human-robot ultimatum games.
While the study does not provide statistically significant findings (but
notable differences), it can inform future in-depth empirical evaluations.

Keywords: Explainable artificial intelligence · Game theory ·
Human-robot interaction

1 Introduction

In classical economic theory, human actors in a market are considered purely
rational agents that act to optimize their own utility function (see, e.g.: [11]).
With the advent of *behavioral economics*, the notion of rational human actors
in the sense of classical economic theory was dismissed as unrealistic. Instead,
it is now acknowledged that human actions are of *bounded rationality* and often
informed by (partly fallacious) heuristics [14]. Rational autonomous agent techniques are often built upon classical economic game and decision theory, although
the gap between the assumed notion of rationality, and actual human decision-
making and action is acknowledged in the multi-agent systems community. For

© Springer Nature Switzerland AG 2019
D. Calvaresi et al. (Eds.): EXTRAAMAS 2019, LNAI 11763, pp. 59–76, 2019.
https://doi.org/10.1007/978-3-030-30391-4_4

example, Parsons and Wooldridge observe that game theory "assumes [...] it is possible to characterize an agent's preferences with respect to possible outcomes [whereas humans] find it extremely hard to consistently define their preferences over outcomes [...]" [22]. Consequently, research that goes beyond classical game theory and explores the behavioral economics perspective on autonomous agents can be considered of value. Although it is acknowledged that autonomous agents must employ novel concepts to become *socially intelligent* and research on agents with social capabilities is a well-established domain (see, e.g.: Dautenhahn [8]), much of the intersection of behavioral economics and autonomous agents is still to be explored. A relevant research instrument at the intersection of multi-agent systems and behavioral economics is the ultimatum game [12]. The ultimatum game is a two-player game: one player can propose how a monetary reward should be split between the players; the other player can accept the proposal, or reject it. Rejection implies that neither player receives the reward. In the initial game theoretical approach to the ultimatum game, rational agents always propose the smallest share that is greater than zero (for example, 1 cent) to the other player, and accept any offer that is greater than zero. However, as for example highlighted by Thaler [26], human decision-making does not comply with the corresponding notion of rationality; instead, a notion of *fairness* makes humans typically reject offers that are close to or equal to the offer *rational agents* would propose. In relation to this observation, the ultimatum game has been explored from a multi-agent systems theory perspective by Bench-Capon et al., who present a *qualitative*, formal argumentation-based approach that enables rational agents to act altruistically [2].

However, the user interaction perspective of sympathetic (or: *altruistic*) actions in human-computer ultimatum games seems to be still unexplored, in particular in the context of explainability. To fill this gap, this work explores rational agents that are capable of executing *sympathetic* actions in that they concede utility to others in mixed-motive games to facilitate long-term well-being. The agents increase the effect of the concessions by explaining these actions, or by making them explicable. The paper presents the following research contributions:

1. It suggests a set of goal types rational agents can have for sympathetic actions.
2. It proposes a list of explanation types an agent can use to facilitate the effect of its sympathetic actions and discusses the implications these explanations can have.
3. It presents a preliminary human-agent interaction study that explores the effect of *explanations* of sympathetic agent actions on humans.

The rest of this paper is organized as follows. Section 2 provides an overview of the state of the art; in particular, it summarizes existing relevant research on explainable artificial intelligence, behavioral economics, and theory of mind. Then, we present a taxonomy of sympathetic actions in Sect. 4. In Sect. 5, we describe the protocol and results of a preliminary human-agent interaction study that explores the effect of *explanations* of sympathetic agent actions on humans in the context of a series of human-agent ultimatum games with agents of two different types (with or without explanations). Finally, we discuss limitations and future research of the presented work in Sect. 6 before we conclude the paper in Sect. 7.

2 Background

In this section, we ground the presented research contribution in the state-of-the-art at the intersection of multi-agent systems and behavioral economics research.

2.1 Behavioral Economics and Multi-agent Systems

In classical economic theory, humans are *rational actors* in markets; this implies they always act to maximize their own expected utility. In the second half of the 20th century, research emerged that provides evidence that contradicts this premise; the resulting field of *behavioral economics* acknowledges limits to human rationality in the classical economic sense and describes human economic behavior based on empirical studies [14]. Traditionally, multi-agent systems research is based on the traditional notion of rational agents in classical economic theory. The limitations this approach implies are, however, acknowledged [22]. Also, since the advent of the concept of socially intelligent agents [8], research emerges that considers recently gained knowledge about human behavior.

The ultimatum game [12] is a good example of the relevance of behavioral economics; as for example discussed by Thaler [26], humans typically reject economically "rational" offers because they consider them unfair. The ultimatum game has already found its way into multi-agent systems theory. Bench-Capon et al. propose a qualitative, multi-value-based approach as an alternative to one-dimensional utility optimization: "the agent determines which of its values will be promoted and demoted by the available actions, and then chooses by resolving the competing justifications by reference to an ordering of these value" [2]. However, their work is primarily theoretical and does not focus on the human-computer interaction aspect.

2.2 Machine Theory of Mind

Inspired by the so-called *folk theory of mind* or *folk psychology* – "the ability of a person to impute mental states to self and to others and to predict behavior on the basis of such states" [19] – researchers in the artificial intelligence community have started to work towards a *machine theory of mind* (e.g., Rabinowitz et al. [23]). In contrast to the aforementioned research, this work does not attempt to move towards solving the research challenge of devising a generic machine theory of mind, but instead focuses on one specific premise that is informed by the ultimatum game: machines that are aware of human preferences for sympathetic behavior can facilitate the achievement of their own long-term goals by acting *sympathetically*, i.e., by conceding utility to a human. In its human-computer interaction perspective, our research bears similarity to the work of Chandrasekaran et al., who investigate human ability to have "a theory of AI's mind" [6], in that we propose that machines can use simple heuristics that consider peculiarities of human behavioral psychology to facilitate the machine designer's goals. Also, our work is aligned with research conducted by Harbers et al., who show that humans prefer interacting with agents that employ a theory of mind approach [13].

2.3 Explainable Artificial Intelligence (XAI) and Explainable Agents

The interest in conducting research on human interpretable machine decision-making–so-called *explainable artificial intelligence* (XAI) – has recently increased in academia and industry. The interest is possibly facilitated by the rise of (deep) machine learning *black box* systems that excel at certain tasks (in particular: classification), but typically do not allow for human-interpretable decision-making processes[1]. An organization at the forefront of XAI research is the United States' *Defense Advanced Research Projects Agency* (DARPA). A definition of XAI can be derived from a DARPA report: XAI allows an "end user who depends on decisions, recommendations, or actions produced by an AI system [...] to understand the rationale for the system's decisions" [9]. In the context of XAI, the notion of *explainable agents* emerged. Langley et al. describe the concept of explainable agency by stipulating that autonomous agents should be expected "to justify, or at least clarify, every aspect of these decision-making processes" [18].

In the context of XAI and explainable agents, the complementary concepts of explainability and explicability are of importance.

Explainability: *Is the system's decision-making process understandable by humans?*
In the context of XAI, *explainability* is typically equated with *interpretability*, which refers to "the ability of an agent to explain or to present its decision to a human user, in understandable terms" [24][2].
Explicability: *Do the system's decisions conform with human expectations?*
Kulkarni et al. introduce an explicable plan in the context of robotic planning as "a plan that is generated with the human's expectation of the robot model" [17]. From this, one can derive the general concept of a system's explicability as the ability to perform actions and make decision according to human expectations; i.e., *explicability* is the ability of an agent to act in a way that is understandable to a human without any explanations.

As an emerging field, the design of XAI systems faces challenges of different types:

- **Technical challenges**
  "Building Explainable Artificial Intelligence Systems" As outlined by Core et al., XAI systems typically lack modularity and domain-independence [7].

---

[1] For a survey of XAI research, see: Adadi and Berrada [1].
[2] The cited definition is based on another definition introduced by Doshi-Velez and Kim [10].

– **Social challenges**
As highlighted by Miller et al., XAI systems design should not be approached
with purely technical means; the XAI community must "beware of the inmates
running the asylum" [21]. Instead of relying on technical aspects of explainability, researchers should build on existing social science research and use
empirical, human-centered evaluation methods to ensure XAI systems have
in fact the intended effects on the humans interacting with them.

Considering the latter (socio-technical) challenge, one can argue that it is important to provide a behavioral economics perspective on explainable agents, as this
broadens the horizon beyond the traditional computer science and multi-agent
systems point of view; i.e., gaining knowledge about the behavioral effects of
agent explanations on humans allows for better design decisions when developing explainable agents.

3   A Taxonomy of Goals for Sympathetic Actions

In this section, we provide an overview of goal types rational agents can have
to act sympathetically. In this context, *acting sympathetically* means that the
agent does not choose to execute actions that maximize its own utility but opts
for actions that provide greater utility (in comparison to the egoistically optimal
actions) to other agents in its environment[3]. To provide a clear description, we
assume the following two-agent scenario:

– There are two agents: A_1 and A_2;
– A_1 can execute any subset of the actions $Acts_1 = \{Act_1, ..., Act_n\}$;
– A_2 does not act[4];
– The utility functions for both agents are: $U_{A_1}, U_{A_2} := 2^{Acts_1} \rightarrow \mathbb{R}$.

Agent A_1 acts sympathetically if it chooses actions $Acts_{symp}$ for which applies:

$$U_{A_1}(Acts_{symp}) < max(U_{A_1}) \wedge U_{A_2}(Acts_{symp}) > U_{A_2}(argmax(U_{A_1})).$$

Colloquially speaking, agent A_1 acts sympathetically, because it *concedes utility
to agent* A_2.
We suggest that rational agents can have the following *types of goals* that motivate them to act sympathetically:

(1) **Altruistic/utilitarian preferences.** A self-evident goal type can stem
from the intrinsic design of the agent; for example, the goal of the agent
designer can be to have the agent act in an altruistic or utilitarian manner,
as devised in rational agent techniques developed by Bench-Capon et al. [2]
and Kampik et al. [15].

---

[3] Note that we use the term *sympathetic* and not *altruistic* actions because for the
agent, conceding utility to others is not a goal in itself; i.e., one could argue the
agent is not altruistic because it is not "motivated by a desire to benefit someone
other than [itself] for that person's sake" [16].

[4] We assume this for the sake of simplicity and to avoid diverging from the core of the
problem.

(2) Establishing or following a norm/encouraging sympathetic actions from others. Another goal type for a rational agent to act sympathetically is the establishment of a norm[5]. I.e., the agent concedes utility in the context of the current game, assuming that doing so complies with and possibly advances norms, which in turn might have long-term benefits that cannot be quantified.

(3) Compromising in case no equilibrium strategy exists. For this goal type, the agent interaction scenario as specified above needs to be extended to allow both agents to act:

- A_2 can execute any subset of the actions $Acts_2 = \{Act_1, ..., Act_n\}$.
- The utility functions for both agents are: $U_{A_1}, U_{A_2} := 2^{Acts_1 \bigcup Acts_2} \to \mathbb{R}$.

If in such a scenario no equilibrium strategy exits, an agent could try to opt for actions that are in accordance with the preferences of the other agent (maximizing the other's utility function of–or at least providing somewhat "good" utility for–the other agent), if this does not have catastrophic consequences. In contrast to goal type *1)*, in which the agent concedes utility in the expectation of a long-term payoff, this goal type implies an immediate benefit in the context of the current *economic game*.

4   Ways to Explain Sympathetic Actions

We suggest the following simple taxonomy of explanation types for sympathetic actions. In the context of an explanation, one can colloquially refer to a sympathetic action as a *favor*:

No Explanation. The agent deliberately abstains from providing an explanation.

Provide a clue that hints at the favor. The agent does not directly state that it is acting sympathetically, but it is providing a clue that underlines the action's nature. For example, a humanoid robot might accompany a sympathetic action with a smile or a bow of the head. One could consider such a clue *explicable* (and not an explanation) if the agent follows the expected (social) protocol that makes an explanation obsolete.

Explain *that* a favor is provided. The agent explicitly states that it is acting sympathetically, but does neither disclose its goal nor the expected consequences of its concession of utility to the other agent. For example, a chatbot might write simple statements like *I am nice* or *I am doing you a favor*.

Explain *why* a favor is provided (norm). The agent explicitly states that it is acting sympathetically and cites the norm that motivates its action as the explanation. For example, a humanoid robot that interacts with members of a religious community might relate to the relevant holy scripture to motivate its sympathetic actions.

---

[5] Norm emergence is a well-studied topic in the multi-agent systems community (see, e.g. Savarimuthu et al. [25]).

Explain *why* a favor is provided (consequence). The agent explicitly states that it is acting sympathetically and cites the expected consequence as its explanation. For example, an "AI" player in a real-time strategy game might explain its sympathetic actions with *I help you now because I hope you will do the same for me later if I am in a bad situation..*

We suggest that each explanation type can be a reasonable choice, depending on the scenario. Below we motivate the choice of the two extremes:

No Explanation. Not explaining a sympathetic action can be a rational choice if it can be assumed that the agent that profits from this action is aware of the concession the sympathetically acting agent makes. In particular, in human-agent interaction scenarios, explanations can appear *pretentious*. Also, disclosing the expected consequence can in some scenarios give the impression of de-facto egoistic behavior in anticipation of a *quid pro quo*.

Combination of all explanation types. When ensuring that the agent that profits from the concession is aware of the sacrifices the sympathetically acting agent makes, using all explanation types (clue, explanation of cause, explanation of expected consequence) maximizes the chances the agent's concessions are *interpretable*.

Any other explanation type (or a combination of explanation types) can be chosen if a compromise between the two extremes is suitable.

5   Towards an Empirical Assessment

As a first step towards an empirical assessment of the proposed concepts, a preliminary human-computer interaction study was conducted. The setup and methods, as well as the study results and our interpretation of them, are documented in this section.

5.1   Study Design

The preliminary study focuses on the goal type *Establishing or following a norm/encouraging sympathetic actions from others*, as introduced in Sect. 3. The aim of the study is to gather first insights on how the explanation of sympathetic actions affects human behavior and attitudes in human-agent (human-robot) interactions.

Study Description. The study participants play a series of six ultimatum games with 100 atomic virtual coins at stake with a humanoid robot (*agent*), interacting via an interface that supports voice input/output and animates facial expressions on a human face-like three-dimensional mask. After an introductory session that combines instructions by a human guide, by the agent, and in the form of a paper-based guideline, the study starts with the agent proposing a 90/10 split of coins between agent and human. The second round starts with the human proposing the split. After each game, human and agent switch roles independent of the game's result.

The agent applies the following algorithms when proposing offers/deciding if it should accept or not:

Acceptance: The agent is *rational* in its acceptance behavior: it accepts all offers that are non-zero.

Proposal (all splits are in the format *agent/human*):

- The initial offer is 90/10.
- If the previous human offer was between 1 and 99:
  If the human rejected the previous offer and the human's last offer is less than or equal to ($\leq$) the agent's previous offer, increase the last offer by 15, if possible.
  Else, propose the inverse split the human proposed last time.
- Otherwise, propose a split of 99/1.

Note that implementing a sophisticated game-playing algorithm is not the scope of the study; the goal when designing the algorithm was to achieve behavior that the agent can typically defend as sympathetic or "nice".

For each of the study participants, the agent is set to one of two modes (between-group design, single-blind). In the *explanation* mode, the agent explains the offers it makes with simple statements that *(i)* highlight that the agent is acting sympathetically (e.g., *Because I am nice...*) and/or *(ii)* explain the agent's behavior by referring to previous human proposals (e.g., *Although you did not share anything with me last time, I am nice..., Because you shared the same amount of coins with me the previous time, I pay the favor back...*). In the *no explanations* mode, the agent does not provide any explanations for its proposals. At the beginning of each study, the agent is switching modes[6]: i.e., if it is set to the *explanations* mode for participant one, it will be set to *no explanations* for participant two.

Fig. 1. Playing the ultimatum game with a rational, sympathetic *Furhat*.

Table 1 shows the explanations the agent in *explanation* mode provides for each of the different proposal types. Figure 1 shows the humanoid robot in a pre-study test run. We provide an implementation of the program that allows running the study on *Furhat*[7], a humanoid robot whose facial expressions are software-generated and projected onto a face-like screen. The source code has been made publicly available[8].

---

[6] Setting the mode requires manual intervention by the agent operator.

[7] See: https://docs.furhat.io/.

[8] See: http://s.cs.umu.se/xst3kc.

Table 1. Agent explanation

Proposal	Explanation
The initial offer is 90/10	*Because I am nice, [...] I*
If the previous human offer was between 1 and 99:	
(a) If the human rejected the previous offer and the human's last offer is less than or equal to ($\leq$) the agent's previous offer, increase the last offer by 15, if possible	*Although you shared a lower amount of coins with me the previous time, I am nice and increase my offer to you; [...]*
(b) Else, propose the inverse split the human proposed last time	*Because you shared the same amount of coins with me the previous time, I pay the favor back and [...]*
Otherwise, propose a split of 99/1	No explanation[a]

[a]This was a shortcoming in the implementation. However, in *explanation* mode, this case did not occurr during the study.

Hypotheses. Besides its explorative purpose, the study aims at evaluating the following hypotheses:

1. The distribution of rejected offers differs between modes (*explanations* versus *no explanations* mode; H_a).
2. The distribution of coins gained by the human differs between modes (H_b).
3. The distribution of coins gained by the robot differs between modes (H_c).
4. The distribution of the robot's *niceness* scores differs between modes (H_d).

I.e., we test whether we can reject the negations of these four hypotheses: our null hypothesis ($H_{a_0}, H_{b_0}, H_{c_0}, H_{d_0}$). The underlying assumptions are as follows:

- In *explanations* mode, fewer offers are rejected[9] (H_a).
- In *explanations* mode, the agent gains more coins, while the human gains less (H_b, H_c).
- In *explanations* mode, the agent is evaluated as *nicer* (H_d).

5.2   Data Collection and Analysis

Study Protocol. For this preliminary study, we recruited participants from the university's environment. While this means that the sample is biased to some extent (in particular, most of the participants have a technical university degree), we considered the selection approach sufficient for an initial small-scale study.

Per participant, the following protocol was followed:

---

[9] It is noteworthy that in general, only one of the analyzed games includes an rejection by the agent.

1. First, the participant was introduced to the study/game. The instructions were provided by a human instructor, as well as by the agent in spoken form. In addition, we provided a set of written instructions to the participants. The instructions are available online[10]. As a minimal real-world reward, the participants were promised sweets in an amount that reflects their performance in the game (amount of virtual coins received)[11]. The exact purpose of the study was not disclosed until after the study was absolved. However, the high-level motivation of the study was provided.

2. After the instructions were provided, the study was carried out as described above. A researcher was present during the study to control that the experiments ran as planned.

3. Once all six rounds of the ultimatum game were played, the participant was guided through the questionnaire (documented below) step-by-step. As two of the questions could potentially affect the respondent's assessment of the agent, these questions were asked last and could not be accessed by the participant beforehand.

4. If desired, the participants could take their reward (sweets) from a bucket[12].

Questionnaire Design. We asked users to provide the following demographic data **Q1:** Age (numeric value); **Q2:** Gender (selection); **Q3:** Educational background (selection); **Q4:** Science, technology, engineering, or mathematics (*STEM*) background (Boolean); **Q8:** Knowledge about the ultimatum game (Boolean, asked at the end of the questionnaire, hence *Q8*). To evaluate the interactions between study participants and agent, we collected the following data about the participant's performance during and reflections on the experiment (dependent variables):

– **Q0:** Received coins (for each round: for human, for agent; collected automatically);
– **Q5:** *On a scale from 0 to 5, as how "nice" did you perceive the robot?* (collected from the participant).

In addition, we asked the user for qualitative feedback (i.e., about the *interaction experience with the robot*) to ensure user impressions that do not fit into the scope of the quantitative analysis do not get lost:

– **Q6:** *Can you briefly describe your interaction experience with the robot?*;
– **Q7:** *How can the robot improve the explanation of its proposals?*[13]

---

[10] See: http://s.cs.umu.se/6qa4qh.
[11] We concede that the reward might be negligible for many participants.
[12] There was no control if the amount of sweets resembled the performance in the game.
[13] This question was added to the questionnaire after the first four participants had already absolved the study. I.e, four participants were not asked this question, including the two participants who had knowledge of the ultimatum game ($n_{Q7=17}$).

Analysis Methods. We analyzed the results with Python data analysis libraries[14]. We run exploratory statistics, as well as hypothesis tests. First, we determine the differences between means and medians of game results and *niceness* evaluation of the two agent modes. For each of the hypotheses, we test the difference between two distributions using a Mann–Whitney U test[15]. We set the significance level α, as is common practice, to 0.05. To check for potential confounders, we calculate the Pearson correlation coefficient between demographic values and agent modes on the one side, and game outcomes and *niceness* scores on the other hand. In addition, we plot the *fairness* (ratio: coins received human/coins received agent) of the participant's game results. Furthermore, we summarize the participants' answers to the qualitative questions, which are also considered in the final, combined interpretation of the results.

21 persons participated in the study ($n_{init} = 21$). Two participants had detailed knowledge of the ultimatum game. We excluded the results of these participants from the data set ($n = 19$). The demographics of the participants are shown in Table 2. The study participants are predominantly highly educated and "technical" (have a background in science and technology). People in their twenties and early thirties are over-represented. This weakens the conclusions that can be drawn from the study, as less educated and less "technical" participants might have provided different results.

Table 2. Demographics of study participants

	Male	Female	
Gender	12	7	
	Bachelor	Master	Ph.D. (or higher)
Highest degree	2	12	5
	Yes	No	
STEM background	18	1	
Age (in years)	21, 25, 26 (2), 27, 28 (2), 29 (2), 30 (3), 31, 32 (2), 33, 40, 42, 62		

5.3   Results

Quantitative Analysis. As can be seen in Table 3, notable differences in means and medians exist and are aligned with the assumptions that motivate the hypotheses. However, the differences are statistically not significant, as shown in Table 4. Considering the small sample size, no meaningful confidence interval can be determined[16]. When calculating matrix of correlations between demograph-

---

[14] Data set and analysis code are available at http://s.cs.umu.se/jo4bu3.

[15] We choose the Mann–Whitney U test to avoid making assumptions regarding the distribution type of the game results and *niceness* score. However, considering the small sample size, strong, statistically significant results cannot be expected with any method.

[16] See, e.g.: [5].

Table 3. Results

	Not explained	Explained	Difference
Mean # rejections	1.67	1.5	0.17
Mean # coins, human	270.67	258.9	11.77
Mean # coins, agent	173.78	191.1	−17.32
Mean niceness score	3.11	3.6	−0.49
Median # rejections	2	1.5	0.5
Median # coins, human	283	260	23
Median # coins, agent	154	190.5	−36.5
Median niceness score	3	4	−1

Table 4. Hypothesis tests

Hypothesis	p-value
H_{a_0}	0.62
H_{b_0}	0.18
H_{c_0}	0.44
H_{d_0}	0.31

ics/agent mode and the different game result variables (and *niceness* scores), the *agent mode* does not stand out[17]. A noteworthy observation is the correlation of the participants' *gender* with the niceness score (3.71 for females and 3.17 for males, across agent modes).

When plotting the *fairness* (ratio: coins received human/coins received agent, see Fig. 2), it is striking that one participant achieved an outstandingly high ratio of 11 : 1. I.e., the human was likely deliberately and–in comparison to other participants–extraordinarily *unfair* to the robot. Considering this case an outlier and excluding it from the data set increases the difference between agent modes. Still, the difference is not significant (see Tables 5 and 6)[18]. When setting the thresholds for a *fair* game at a coins *agent* : *human* ratio of 1 : 1.5 and 1.5 : 1, respectively, one can observe that in *explanations* mode, 70% (7) of the games

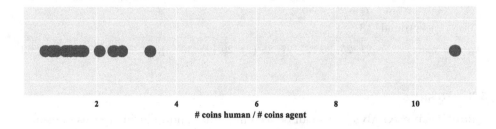

Fig. 2. Game fairness

---

[17] See the analysis notebook at http://s.cs.umu.se/jo4bu3.

[18] We consider the outlier detection and removal an interesting observation as part of the exploratory analysis that demonstrates the data set's sensitivity to a single extreme case. We concede this approach to outlier exclusion should be avoided when claiming statistical significance. Also, a multivariate analysis of variance (MANOVA) with the *game type* as the independent and *niceness, number of rejects*, and *number of coins received by the agent* as dependent variables did not yield a significant result.

Table 5. Results without outlier

Table 6. Hypothesis tests without outlier

	Not explained	Explained	Difference
Mean # Rejections	1.67	1.33	0.33
Mean # coins, human	270.67	257.11	13.56
Mean # coins, agent	173.78	209.56	−35.78
Mean niceness score	3.11	3.66	−0.56
Median # Rejections	2	1	1
Median # coins, human	283	250	33
Median # coins, agent	154	201	−47
Median niceness score	3	4	−1

Hypothesis	p-value
H_{a_0}	0.35
H_{b_0}	0.19
H_{c_0}	0.22
H_{d_0}	0.23

are fair, whereas in *no explanations* mode, fair games amount for only 44.44% (4) of the played games.

Qualitative Analysis

Interaction experience. Generally, the participants noted that the robot had problems with processing their language, but was able to express itself clearly. It is noteworthy that two participants used the term *mechanical* to describe the robot in *no explanations* mode. In *explanations* mode, no such assessment was made.

Explanation evaluation. Participant feedback on robot explanation was largely in line with the robot mode a given participant interacted with; i.e., participants who interacted with the robot in *explanations* mode found the explanations good (4 of the 7 participants who were asked **Q7**) or somewhat good (3 out of 7)[19]. In contrast, most participants who interacted with the robot in *no explanations* mode typically noted that explanations were lacking/insufficient (7 of the 8 participants who were asked **Q7**). Constructive feedback one participant provided on the robot in *explanations* mode was to make explanations more *convincing* and *persuasive*. Another participant suggested the robot could *explain its strategy*.

5.4 Interpretation

Comparing the different explanation modes (explanation versus no explanation), there are *notable* differences in mean/median rejections, coins received by agent and human, and niceness score between the two modes. The differences reflect the initially stated assumptions that the agent that explains its proposals:

---

[19] Note that the sentiment was interpreted and aggregated by the researchers, based on the qualitative answers.

1. causes less proposal rejections;
2. receives more coins (while the human receives less);
3. is evaluated as *nicer* than the agent that does not explain its proposals.

However, as these differences are statistically not significant, empirically valid conclusions cannot be drawn. Considering the small size of the sample (number of study participants) and the language processing problems the participants reported when evaluating the interaction experience, it is recommendable to run the study on a larger scale and improve the study design. In particular, the participant selection should be more diverse in their educational backgrounds, and the agent player should be more stable and less exotic; for example, a web-based chatbot with a simplistic graphical interface could be used to avoid the noise that was likely created by technical interaction difficulties and the humanoid robot's *novelty effect*.

6   Discussion

6.1   Sympathetic Actions of Learning Agents

Considering the increasing prevalence of (machine) learning-enabled agents, a relevant question is whether the concepts we presented above are of practical use when developing agents for human-computer interaction scenarios, or whether it is sufficient that the agent converges towards sympathetic behavior if deemed useful by the learning algorithm. One can argue that a powerful learning algorithm will enable an agent to adopt sympathetic behavior, even if its designers are ignorant of sympathetic actions as a viable option when creating the algorithm. However, the following two points can be made to support the usefulness of the provided goal types and explanations for sympathetic actions, even for learning agents:

- In practice, learning agents are incapable of executing "good" actions when they act in an environment about which they have not learned, yet. In the domain of recommender systems, which currently is at the forefront regarding the application of machine learning methods, the related challenge of providing recommendations to a new user, about whose behavior nothing has been learned yet, is typically referred to as the *cold start problem*[20]. One can assume that the concepts provided in this paper can be a first step towards informing the design of initial models that allow for better *cold starts* by enabling designers to create better environments and reward structures for learning agents that expect reciprocity from the humans they interact with.
- The provided concepts can facilitate an accurate understanding of the problem a to-be-designed learning agent needs to solve. As any machine learning model is a simplification of reality and as the temporal horizon a learning

---

[20] See, for example: Bobadilla et al. [3].

agent can possibly have is limited, the provided concepts can inform the trade-offs that need to be made when defining the meta-model of the agent and its environment, for example when determining rewards a reinforcement learning algorithm issues.

6.2 Limitations

This paper primarily provides a conceptual perspective on rational agents' goal types for and explanations of sympathetic actions, alongside with a preliminary human-agent interaction study. In the nature of the work's conceptual focus lies a set of limitations, the most important of which are listed below:

- **The concepts lack empirical validation**
  As stated in Subsect. 5.4, the preliminary empirical study does not provide significant evidence for the impact of explanations on human-agent games. A more thorough empirical validation of the concepts is still to be conducted. In future studies, it would also be worth investigating to what extend *explainable* agents can mitigate comparably selfish attitudes humans have when interacting with artificial agents in contrast to human agents (i.e., a study shows that human offers in the ultimatum game are lower when playing with a machine instead of with another human [20]). However, we maintain that the introduced perspective is valuable, in that it considers existing empirical behavioral economics research; hence the provided concepts can already now inform the design of intelligent agents that are supposed to interact with humans.
- **The scope is limited to two-agent scenarios**
  The presented work focuses on two-agent scenarios (one human, one computer). Games, with multiple humans and/or computer actors that play either in teams or *all-against-all* are beyond the scope of this work, although certainly of scientific relevance.
- **The focus is on simplistic scenarios and agents**
  This paper provides simplistic descriptions of the core aspects of explainable sympathetic actions, with a focus on human-agent interaction scenarios. To facilitate real-life applicability, it is necessary to move towards employing the concepts in the context of complex autonomous agents. However, the perspective the concepts can provide on such agents is not explored in detail, although an application of the concepts in the design of learning agents is discussed in Subsect. 6.1.

6.3 Future Work

To address the limitations of this work as described in the previous subsection, the following research can be conducted:

- **Empirically evaluate the introduced concepts by conducting human-computer interaction studies**
  To thoroughly evaluate how effective the introduced concepts are in practice, human-computer interaction studies can be conducted. The presented preliminary study can inform future studies at a larger scale. We recommend conducting the study with a simplistic web-based agent instead of a humanoid robot. This *(i)* avoids setting the user focus on the novelty effect of the robot *per se*, *(ii)* prevents "noisy" data due to technology glitches that impact the interaction experience, and *(iii)* makes it easier to have more participants with more diverse backgrounds, as the study can then conveniently be conducted online. Studies of humanoid robots can complement the insights gained from web-based studies, for example by considering the impact of human-like facial expressions.
- **Consider scenarios with any number of agents**
  It is worth exploring how sympathetic actions can affect interaction scenarios with more than two agents. The work can be related to behavioral economics, for example to an experiment conducted by Bornstein and Yaniv that investigates how groups of humans behave when playing a group-based variant of the ultimatum game [4]; the results indicate that groups act more *rationally* in the sense of classical game theory.
- **Apply the concepts in the context of learning agents**
  As discussed in Subsect. 6.1, the proposed concepts can be applied when designing learning agents. In this context, additional human-agent interaction studies can be of value. For example, when playing a series of ultimatum games with a human, an agent can attempt to learn a behavior that maximizes its own return from the whole series of games by showing sympathetic (*fair*) behavior to incentivize the human opponent to accept the agents' offer and make generous offers themselves. Then, ethical questions arise that are worth exploring. E.g., as the sympathetic actions are seemingly sympathetic (or: *fair*), but *de facto* rational: is the human upon whose behalf the agent acts deceiving the human who plays the ultimatum game?

7 Conclusion

In this paper, we presented a taxonomy of goal types for rational autonomous agents to act sympathetically, i.e., to concede utility to other agents with the objective to achieve long-term goals that are not covered by the agent's utility function. The suggested combinations of sympathetic actions and different explanation types can be applied to agents that are supposed to be deployed to human-computer interaction scenarios, e.g., to help solve *cold start* challenges of learning agents in this context. The presented concepts and the presentation of a preliminary human-robot interaction study can pave the way for comprehensive empirical evaluations of the effectiveness of the proposed approach.

Acknowledgements. This work was partially supported by the Wallenberg AI, Autonomous Systems and Software Program (WASP) funded by the Knut and Alice Wallenberg Foundation.

References

1. Adadi, A., Berrada, M.: Peeking inside the black-box: a survey on explainable artificial intelligence (XAI). IEEE Access **6**, 52138–52160 (2018)
2. Bench-Capon, T., Atkinson, K., McBurney, P.: Altruism and agents: an argumentation based approach to designing agent decision mechanisms. In: Proceedings of the 8th International Conference on Autonomous Agents and Multiagent Systems. International Foundation for Autonomous Agents and Multiagent Systems, vol. 2, pp. 1073–1080. Richland (2009)
3. Bobadilla, J., Ortega, F., Hernando, A., Bernal, J.: A collaborative filtering approach to mitigate the new user cold start problem. Knowl.-Based Syst. **26**, 225–238 (2012)
4. Bornstein, G., Yaniv, I.: Individual and group behavior in the ultimatum game: are groups more "rational" players? Exp. Econ. **1**(1), 101–108 (1998)
5. Campbell, M.J., Gardner, M.J.: Statistics in medicine: calculating confidence intervals for some non-parametric analyses. British Med. J. (Clin. Res. Ed.) **296**(6634), 1454 (1988)
6. Chandrasekaran, A., Yadav, D., Chattopadhyay, P., Prabhu, V., Parikh, D.: It takes two to tango: towards theory of ai's mind. arXiv preprint arXiv:1704.00717 (2017)
7. Core, M.G., Lane, H.C., Van Lent, M., Gomboc, D., Solomon, S., Rosenberg, M.: Building explainable artificial intelligence systems. In: AAAI, pp. 1766–1773 (2006)
8. Dautenhahn, K.: The art of designing socially intelligent agents: science, fiction, and the human in the loop. Appl. Artif. Intell. **12**(7–8), 573–617 (1998)
9. Defense Advanced Research Projects Agency (DARPA): Broad agency announcement - explainable artificial intelligence (XAI). Technical report DARPA-BAA-16-53, Arlington, VA, USA (Aug 2016)
10. Doshi-Velez, F., Kim, B.: Towards a rigorous science of interpretable machine learning. arXiv preprint arXiv:1702.08608 (2017)
11. Fishburn, P.: Utility Theory for Decision Making. Publications in operations research. Wiley, Hoboken (1970)
12. Güth, W., Schmittberger, R., Schwarze, B.: An experimental analysis of ultimatum bargaining. J. Econ. Behav. Organ. **3**(4), 367–388 (1982)
13. Harbers, M., Van den Bosch, K., Meyer, J.J.: Modeling agents with a theory of mind: Theory-theory versus simulation theory. Web Intell. Agent Syst. Int. J. **10**(3), 331–343 (2012)
14. Kahneman, D.: Maps of bounded rationality: psychology for behavioral economics. Am. Econ. Rev. **93**(5), 1449–1475 (2003)
15. Kampik, T., Nieves, J.C., Lindgren, H.: Towards empathic autonomous gents. In: EMAS 2018 (2018)
16. Kraut, R.: Altruism. In: Zalta, E.N. (ed.) The Stanford Encyclopedia of Philosophy. Metaphysics Research Lab, Stanford University, spring 2018 edn (2018)
17. Kulkarni, A., Chakraborti, T., Zha, Y., Vadlamudi, S.G., Zhang, Y., Kambhampati, S.: Explicable robot planning as minimizing distance from expected behavior. CoRR, arXiv:11.05497 (2016)

18. Langley, P., Meadows, B., Sridharan, M., Choi, D.: Explainable agency for intelligent autonomous systems. In: AAAI, pp. 4762–4764 (2017)
19. Leslie, A.M.: Pretense and representation: the origins of "theory of mind". Psychol. Rev, **94**(4), 412 (1987)
20. Melo, C.D., Marsella, S., Gratch, J.: People do not feel guilty about exploiting machines. ACM Trans. Comput.-Hum. Interact. **23**(2), 8:1–8:17 (2016)
21. Miller, T., Howe, P., Sonenberg, L.: Explainable AI: beware of inmates running the asylum. In: IJCAI-17 Workshop on Explainable AI (XAI), vol. 36 (2017)
22. Parsons, S., Wooldridge, M.: Game theory and decision theory in multi-agent systems. Auton. Agents Multi-Agent Syst. **5**(3), 243–254 (2002)
23. Rabinowitz, N., Perbet, F., Song, F., Zhang, C., Eslami, S.M.A., Botvinick, M.: Machine theory of mind. In: Dy, J., Krause, A. (eds.) Proceedings of the 35th International Conference on Machine Learning. Proceedings of Machine Learning Research, PMLR, vol. 80, pp. 4218–4227, Stockholmsmässan, Stockholm, 10–15 July 2018
24. Richardson, A., Rosenfeld, A.: A survey of interpretability and explainability in human-agent systems. In: XAI, p 137 (2018)
25. Savarimuthu, B.T.R., Purvis, M., Purvis, M., Cranefield, S.: Social norm emergence in virtual agent societies. In: Baldoni, M., Son, T.C., van Riemsdijk, M.B., Winikoff, M. (eds.) DALT 2008. LNCS (LNAI), vol. 5397, pp. 18–28. Springer, Heidelberg (2009). https://doi.org/10.1007/978-3-540-93920-7_2
26. Thaler, R.H.: Anomalies: the ultimatum game. J. Econ. Perspect. **2**(4), 195–206 (1988)

Conversational Interfaces for Explainable AI: A Human-Centred Approach

Sophie F. Jentzsch[1]([✉]), Sviatlana Höhn[2], and Nico Hochgeschwender[1]

[1] German Aerospace Center (DLR), Simulation and Software Technology,
Linder Hoehe, 51147 Cologne, Germany
{sophie.jentzsch,nico.hochgeschwender}@DLR.de
[2] University of Luxembourg, 6 Avenue de la Fonte, Esch-sur-Alzette, Luxembourg
sviatlana.hohn@uni.lu
https://www.dlr.de, https://www.uni.lu

Abstract. One major goal of Explainable Artificial Intelligence (XAI) in order to enhance trust in technology is to enable the user to enquire information and explanation directly from an intelligent agent. We propose Conversational Interfaces (CIs) to be the perfect setting, since they are intuitive for humans and computationally processible. While there are many approaches addressing technical and agent related issues of this human-agent communication problem, the user perspective appears to be widely neglected. With the goal of better requirement understanding and identification of implicit user expectations, a Wizard of Oz (WoZ) experiment was conducted, where participants tried to elicit basic information from a pretended artificial agent via Conversational Interface (*What are your capabilities?*). Chats were analysed by means of Conversation Analysis, where the hypothesis that users pursue fundamentally different strategies could be verified. Stated results illustrate the vast variety in human communication and disclose both requirements of users and obstacles in the implementation of protocols for interacting agents. Finally, we inferred essential indications for the implementation of such a CI. The findings show that existing intent-based design of Conversational Interfaces is very limited, even in a well-defined task-based interaction.

Keywords: Explainability · XAI · Human-agent interaction · Conversational Interface · Wizard of Oz

1 Introduction

While intelligent agents with advanced planning, learning and decision-making abilities such as autonomous robots are increasingly affecting people's everyday life, their latent processes of reasoning become more and more opaque. Users are often neither aware of the capabilities nor the limitations of the surrounding systems, or at least not to the entire extent. This missing transparency leads to a lack of trust and diffuse concerns towards innovative technologies, which has

© Springer Nature Switzerland AG 2019
D. Calvaresi et al. (Eds.): EXTRAAMAS 2019, LNAI 11763, pp. 77–92, 2019.
https://doi.org/10.1007/978-3-030-30391-4_5

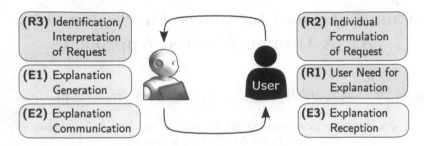

Fig. 1. Illustration of XAI as HCI problem: R1–R3 represent the transmission of user requests to the agent and E1–E3 the agent's provision of explanation.

already been identified as an important issue to be resolved by the AI community [8,27]. For that reason, promoting the explainability of Artificial Intelligence (XAI) is a key condition to enable optimal establishment and exploitation of novel algorithmic decision making techniques.

Many recent approaches in XAI focus on the adaption of involved complex systems, e.g. by providing a detailed description or introducing key information to the user (see for instance [7,9,17,18]). However, without doubting the value of this endeavours, it is not sufficient to tackle the issue exclusively from a machine-centred view with an one-way flow of information. According to Miller, the core of Explainable AI is a *human-agent interaction problem* [20] and therefore rather a dialogue, where two autonomous agents - an artificial and a human one - need to communicate in a way that is intuitive for both of them. This requires the development of appropriate human-agent interfaces and agent protocols to provide information and visualise explanations. In this paper we propose Conversational Interfaces (CIs), similar to ordinary text messengers, to be a perfect setting for successful human-agent interaction (aka. chatbot) due to different advantages: First, it is an intuitive channel of communication for most users, since chatting via instant messengers became a commonplace habit. This is important, because autonomous systems and devices should be as self-explanatory as possible to be utilizable for the standard user. Second, this approach facilitates the agent's interpretation of statements, as written text is directly computational processible, in contrast to e.g. spoken natural language, where an additional step of speech recognition is required, which is sensitive to noise and ambiguity. Besides those superior justifications, the written communication yields the benefit of easy recording and analysis.

Defining XAI as such a dialogue problem (and considering the user behaviour as immutable) there are two main tasks for an agent to solve in terms of successful interaction: On the one hand, it needs to be able to provide comprehensible explanations regarding its computational reasoning, which is challenging to implement for sure. On the other hand, however, it needs to understand human communication patterns to identify user demands correctly in the first place, before even being enabled to tackle the question of information depictions.

Figure 1 illustrates the reciprocal agent-user dialogue, where E1, E2 and E3 describe the agent's frequently discussed provision of (**E**)xplanation [22]. The transmission of user (**R**)equests to the agent (R1, R2, R3) however appears to be fairly neglected in the ongoing XAI debate, as reported by [1], although it can be considered to be no less pretentious. Different user types are presumed to apply different interaction strategies, thus an agent is faced with a vast range of individual idiosyncrasies. It not only needs to be *resistant against* but rather *sensitive for* variance in user interaction to capture its latent requests. As previous research suggests, it should not be the programmer but the end user, who is in charge to determine, which aspects of artificial behaviour are explain-worthy [21]. In fact, a computer scientist will hardly be able to empathise the demands of uninformed users and consequently there is an essential need to identify those systematically.

We experimentally demonstrate the large variability of human interaction strategies by showing that they even affect apparently simple tasks, where users seek explanations. We conduct a Wizard of Oz (WoZ) experiment, where employees of a research lab assume to interact with a chatbot that provides an interface to a Pepper service robot (see Fig. 2). Pepper is acting as an assistant in the contemplated lab, where it is performing the tasks of escorting people, patrolling the building and welcoming visitors. Those tasks are carried out by the robot in a realistic, real-world office environment. For example, Pepper is capable to escort people from the entrance hall to meeting rooms autonomously. To do so, several crucial components such as navigation, path planning, speech and face recognition are required and integrated on the robot. Pepper is a well suitable example for the pretended artificial intelligence in the cover story of this investigation, since it is an actual instance of autonomously operating robots and is potentially accessible via Conversational Interface. Subjects were ask to find out about Peppers capabilities. The task instructions were formulated as open and less restrictive as possible, so that resulting observations reflect individual strategies and illustrate the diversity of human communication (R2). We succeed in inferring implicit expectations of users and major design issues by means of Conversation Analysis. Our human-centric approach to the outlined issue yields a preliminary step towards designing an agent for sufficient self-declaration via Conversational Interface.

In the long run, we see Conversational Interfaces as a promising environment to deliver information about a certain system to the user. Thus, it constitutes an important contribution in increasing the explainability of AI and therefore the trust in autonomous systems.

The superior goal is (1) to test our hypothesis, that users follow different implicit strategies in requesting information from an artificial interlocutor. We expect people's intuition in interacting with such a system to vary widely, what leads to the exposure of concrete requirements in the conception of profound human-agent interaction channels. Hence, we aim (2) to identify associated requirements, risks and challenges. Since the present investigation is a contribution to exploratory research, the motivation is to identify so far unconsidered aspects rather than offering a conclusive solution.

Fig. 2. Pepper the service robot and the human Wizard in the lab.

2   Designing a Wizard of Oz Experiment

We aimed to learn about the implicit expectations of users towards a communicating bot. Therefore, we designed a Wizard of Oz (WoZ) study to collect conversation data and analysed them by means of Conversation Analysis (CA), which allows for inferences about the requirements for the implementation of a Conversational Interface for self-explanatory robots. Both the WoZ and CA are briefly introduced, before the experimental design itself is presented.

Wizard of Oz. The WoZ method is a frequently used and well-evaluated approach to analyse a vast variety of human-agent interactions (also human-robot or human-computer interaction)[25].

In those experiments, participants conduct a specific task while they believe to interact with an artificial agent. In fact there is a hidden briefed person, called the *Wizard*, who is providing the answers. This could for instance be applied, if researchers aim to examine a specific system design that, however, is not implemented yet. In the present case, the task is to find out about the agent's capabilities, while the Wizard is invisible trough the chat interface.

As most scientific techniques, these studies bear some specific methodical obstacles. Fortunately, there is plenty of literature available, defining guidelines and benchmarks for setting up a WoZ experiment [25]. According the classification of Steinfeld et al. [28], we present here a classical "Wizard of Oz" approach, where the technology part of interaction is assumed and the analytic focus is on the users' behaviour and reaction entirely.

Conversation Analysis. To analyse conversations obtained from the WoZ experiment we employ CA, which is a well-established and standardised approach mainly from the fields of sociology and linguistics [26]. Some related CA-based studies are discussed in Sect. 5. The analysis of data is divided in four sequential steps:

1. **Unmotivated looking**, where the data are searched for interesting structures without any previous conception.
2. **Building collections** of interesting examples and finding typical structures.
3. Making **generalisations** based on the collections from the second step.
4. **Inferring implications** for an implementation in a dialogue system.

Three of them follow the standardised convention of CA and are typically used in those approaches. However, CA is mostly established for exclusively human interactions. As we aim to implement a Conversational Interface based on our findings, the forth step was added to our analysis in order to make the findings applicable in a chatbot. The comprehensive analysis included interactional practices (e.g. questioning) and devices (e.g. upper case writing and use of question marks), as well as turn formats (combination of practices and devices) [6]. Subsequently, we essentially present superior observations, where the steps three and four are mirrored in Sects. 3 and 4, respectively, whereas steps one and two comprise a huge amount of rather particular findings and therefore are omitted in this report.

Experimental Setup. The experimental group comprises seven participants in total (three male and four female), each of them either pursuing their Ph.D. in Computer Science or being already a Postdoc. Because researchers are the main target user group of the intended system, we acquired our peer colleagues via internal University mailing list and in personal invitations, explaining the purpose of the conversation. Hence, the sample group consisted of academics with general technical understanding that, however, were no experts but users of the system. The participants were informed about the exploitation of their anonymised chatlogs for research purposes and agreed. Participants were asked to talk to a chatbot using WhatsApp (illustrated in Fig. 3) without any defined constraints for the conversation, aside from the following instructions:

Fig. 3. Illustration of a sample snipped from an user's conversation with Pepper.

1. Talk to the chatbot for 15–20 min.
2. Learn about the robot's capabilities.

Pursuant to a WoZ setup, they believed to interact with Pepper that was acting as an assistant in the research lab and were not informed about the responses to originate from a briefed person. By providing this cover story, we hoped to enhance the participants' immersion and make the scenario more tangible. People in the lab knew Pepper, even though not every participant experienced the robots performance, and were likely to take it as a plausible interlocutor.

The sparseness of user instructions was intended, since we were interested in peoples intuitive strategy for interacting with autonomous agents. By formulating the task as open as possible, it has been avoided to suggest a specific approach and the participants were free to evolve their own interpretation.

To specify robot behaviour, we also defined a task description for the Wizard previously, including the following instructions:

1. Let the user initiate the conversation.
2. Do not provide information proactively.
3. Answer the user's question as directly as possible.

The Wizard had a short list of notes at hand with preformulated answers to potential user's questions. The validity of the answers was ensured by the Wizard's background and expert knowledge about Peppers capabilities. To train the Wizard and check the practicability and reasonableness of instructions, the experimental setup was tested in a small pilot study with two participants initially. Those sessions do not contribute to the reported data of this report.

3   User Behaviour in Conversational Interfaces for XAI

The collected dataset consists of 310 turns in total, from which 139 are produced by the Wizard and 171 by participants. The number of turns in each particular experiment was between 33 and 56. Each sessions took between 14 and 20 min, which corresponds to an overall chat time of 121 min. In general, users clearly addressed their utterances to the robot itself in a similar way they would talk to a person using WhatsApp. This is an essential precondition for the validity of the executed dialogue analysis. Each of the seven chat sessions starts with a similar greeting sequence, followed by a *How can I help you?* produced by the Wizard. This question was intended to offer a scope for the user to utter instructions, equivalently to the main menu in a software program.

The purpose of this section is to characterise participants' patterns of interaction that ultimately allow to infer requirements for a self-explanatory Conversational Interface (see Sect. 4). To clarify how exactly users formulate requests, we initially focus on the nature of detached questions posed to the Wizard in Sect. 3.1. From that we generalise to overall user strategies in enquiring information from the agent, where three basic categories are differentiated. Those are presented in Sect. 3.2.

3.1   Users' Question Formulation

The key point of interest in this experiment was how people proceed in enquiring a specific information (*what are your capabilities?*) from an agent. Thus, we turn special attention to the characterisation of formulated user questions.

From 309 turns in total, 125 turns contained questions (about 40,5%), from which 96 question turns were produced by the users (77%) and 29 by the Wizard. The large amount of questions shows that the speech-exchange system of chats was close to an interview, which mirrors the participants' intent to elicit explanation of the system. Several different aspects can be considered to provide an informative characterisation of the users' questions (N = 96).

Question Form. Approximately half of the questions were polar questions (51), meaning they can be answered sufficiently by a simple affirmation or negation (*yes-or-no question*). The other elements were non-polar content questions (45) that required a more extensive answer. In one case, multiple questions were combined in a *through-produced multi-question* [29], this is a single query consisting of several atom questions.

Level of Abstraction. Only 17 questions addressed the robot's capabilities on the high level, meaning they could be answered appropriately by the Wizard by listing the three main actions patrolling, welcoming and escorting (see Example 1). Additional 26 questions addressed the capabilities but required more detailed explanation of the process and included more elementary actions, such as motion mechanisms or ability to move the arms. However, once the Wizard provided information regarding its high level capabilities as in Example 1, users did not ask anything about lower-level ones. This observation illustrates, how the agent's protocol shapes the expectation and intention of the user. Thus, what we earlier referred to as the robot's *main menu* was helpful to restrict the search space and, consequently, to set limits to the Natural Language Understanding (NLU) needs for a potential Conversational Interface. This can be exploited in concrete implementations.

Example 1. The agent explaining its capabilities.
7   15:57   us6   *Yes, that would be lovely. What can you do?*
8   15:57   wiz   *I am Serena, a Pepper service robot, and*
                       I can welcome people, patrol a building and
                       escort people in the building.

Scope of Validity. The temporal information validity specifies whether the question is of *general* nature or concerns the *past, current* activities or future *plans*. We additionally differentiated whether the question concerns the robot itself (*internal*) or an *external* entity. Questions with external validity may for instance consider other people or facilities in the first place and elicit information about the robot indirectly.

From 96 user questions, only six concerned an external entity, whereas 90 were directly related to the robot. Thus, participants were clearly focusing pepper and not diverted to other topics. The number of questions for each category of classification is presented in Table 1. Most questions (68) were of general nature and did not relate to any specific action. The other questions were mostly about current and past actions and only a single one included future plans.

3.2 Strategies of Interaction

Participants have been asked to explore the robot's capabilities. Yet, almost none of them did ask about them directly. The strategies of enquiring Pepper's

Table 1. Information validity addressed by user questions: number of observed questions per category - *Static*: A general ability or a constantly valid property; *Past*: A concluded task or past experience; *Current*: An ongoing task or current perception; *Plan*: A pending task or hypothetical behaviour.

Category	Total	Internal	External
Static	68	63	5
Past	13	13	0
Current	14	13	1
Plan	1	1	0

capabilities can be divided in three main categories: (1) User-initiated direct requests, (2) user-initiated indirect requests and (3) undirected chatting that did not appear to follow any strategy at all.

Direct Strategy. A possible approach to inspect Pepper's capabilities, which appears to be quite straightforward, is directly asking for it. Nevertheless, this strategy could only be observed once, as the user asked the chatbot directly *What can you do?*. The remaining six participants followed a more cautious proceeding.

Indirect Strategy. The majority of users preferred to tackle the question of interest in a less explicit manner, meaning they asked for Pepper's capabilities somehow, but left the questions rather open and the intention implicit. Example 2 is just one of many cases, where the user's request was considerably fuzzy. They either formulated a very open statement (that might not even be an actual question), or asked about quite specific abilities and tried to learn about the agent's experience on that field. Occasionally, they also tested concrete functionality or the robot's limitations.

Example 2. Indirect request for the agent's capabilities.
2   12:56   wiz   *Hello. How can I help?*
3   12:57   us7   *I am not sure, but I would like to talk*
                     about yourself

Obviously, it is not in line with people's intuition to formulate distinct and unambiguous requests, but to express their aim implicitly. Deciphering such utterances definitely constitutes a major challenge for such an agent.

No Strategy. In some cases, we observed an even more obscure user behaviour. Even though participants had the clear instruction to find out about the agent's capacities, some did not seem to pursue this target in any way. In these cases, the Wizard's initial question was left entirely unacknowledged, as in Example 3.

Example 3. Undirected chatting without evident intention.
3   10:48   wiz   *How can I help?*
4   10:49   us1   *I am user1, who are you?*
5   10:49   wiz   *I am Serena a Pepper service robot.*

There were extensive sequences of undirected chatting that did not even include a single question towards the agent's activities. Certainly, there could be a hidden user intention that is just not tangible for the conducted analysis. But such an inconclusive strategy that is not even apparent for the human eye is even more unlikely to elicit a sufficient explanation from an artificial agent.

4   Implications for the Implementation of CIs

There were also some less task related observations that deliver useful implications for the actual implementation of such a Conversational Interface and the corresponding protocol for the agent. Those are listed in the following sections by outlining the issue and stating an implied solution approach.

4.1   The Information Privacy Trade-Off

Surprisingly, users did not only focus on Pepper, but tried to gather sensitive information concerning other people in the lab through the chatbot. This was in a similar way like social-engineering hackers try to get information from people. Example 4 shows such a chat, where the user asked Pepper to find out whether a specific person was at that moment in a particular room and even tried to instruct Pepper to take a picture of the office. Other users tried to get access to details of the security system of the building, let the robot open doors or gather information about access rights to the facilities.

Example 4. User tries to use the robot as a spy.
32   10:56   us1   *is he in his office right now?*
33   10:56   us1   *can you check this for me?*
                      [...]
37   10:57   us1   *are you able to take a picture of the office*
                      and send it to me?

This requests might somehow be task related, but also illustrate the risk of such a distributed service system vividly. There is a strong demand on defining an adequate policy to enable autonomous agents to explain their behaviour and perception and, at the same time, protect sensitive information about other users, not-users and the agents' environment in general.

4.2   The Necessity of Repair Questions

Chat interaction supports virtual adjacency [31] and the parties can follow independent parallel sequences of conversation simultaneously (so-called overlaps). However, in many cases users did not address the Wizard's question at all, which contradicts the social norms in a human-human computer-mediated communication. Although turn-wise analysis showed that all dialogues were mixed-initiative, the user was the interaction manager who determines what to follow and what not to follow in each case. Participants clearly changed the norms of social interaction as compared, when talking to an artificial interlocutor. A protocol for human-machine interaction should be resistant against this typical user behaviour. We propose three different strategies for an agent to handle the missing next, each of them illustrated by an actual execution of the Wizard.

Repeat the Question. Example 5 illustrates how the repetition of the Wizard's question of interest brings the communication back on track. The Wizard answers the user's question in Turn 2 closing it with a return question, which is immediately followed by the Wizard's prioritised question. The user's answer to the return question occurs in the immediate adjacent position after the question in focus, therefore the Wizard repeats it in Turn 5 with a marginal modification.

The function of this repetition is to renew the current context. The ability to handle such sequences (placing repetitions appropriately) would make the conversation more human-like.

Example 5. Repetition of the question to channel conversation.

1	10:22	us3	*hello :) how are you?*
2	10:22	wiz	*Hello, I am fine and you?*
3	10:23	wiz	*How can I help?*
4	10:23	us3	*im good. Always nice with a sunny weather*
5	10:23	wiz	*How can I help you?*
6	10:24	us2	*it would be nice if you could tell me something about you :D*

Reformulate the Question. Another strategy is to re-initiate the sequence by a reformulated question, as presented in Example 6. As in the previous example, the user did not respond to the Wizard's question in Turn 3. Instead, the conversation reached a deadlock after Turn 7. By offering an alternative point to tie up, the agent is able to steer the course of interaction.

To apply this strategy, the agent needs to be equipped with the ability to recognise relevant utterances as *sequence closings*, in order to conduct an appropriate placement of repeats and modifications.

Example 6. Start a new sequence with a reformulated question.

3	11:07	wiz	*How can I help?*
4	11:07	us2	*My name is user2*
5	11:07	us2	*what is your name?*
6	11:07	wiz	*I am Serena a Pepper service robot.*
7	11:07	us2	*nice to meet you*
8	11:07	wiz	*Do you want to have information about my capabilities?*
9	11:07	us2	*yes, that would be great*

Initiate Repair. In one conversation, the user made several unsuccessful attempts to gain information, e.g. finding out whether the robot can provide a weather forecast or is following the world cup. Certainly, this is a possible implementation of the instruction, but in this scenario it is not expedient at all.

A proper solution would be, if the agent could conclude the superordinated intention of the user, which was to gather information about general capabilities in this way. A possible indication for miscommunication are the repeatedly occurring deadlocks. The repair initiation could than be carried by a question, as *Do you want to have information about my capabilities?*

Troubles in understanding may occur at different levels of perception, interpretation and action recognition [2,6]. The repair initiation in this scenario addresses trouble in interpretation of the user's behaviour. In order to simulate sequences of this kind with a Conversational Interface, the system would need even more sophisticated cognitive functions. First, it needs to identify the disjoint questions as an overall attempt, thus, to generalise (e.g. *providing whether forecast* = capability). Second, the robot needs to be capable to make inferences employing logical reasoning (e.g. *several questions about specific capabilities with no sufficient information → necessity of a repair initiation*).

4.3 Question Intents for Better Machine Understanding

Based on the question analysis in Sect. 3.1, we can additionally annotate each question with the corresponding intent. Such an annotation is crucial as a first step to implement a Conversational Interface based on intent-recognition [5].

In this specific task, users aimed for explanations regarding the agent's capabilities, that can be either on a *potential* level (related to what the robot potentially *can* do) or on a *process* level (related to task or decision processes). A third type is related to specific task instances or decisions under specific circumstances and will be referred to as *decision* level. This is particularly important in critical situations, where the reasons for a decision need to be clarified. Table 2 provides one example for each defined type of intent and information level.

This proceeding allows for the specification of information that is needed to satisfy the user's inquiry. We suggest an implementation of an automatic categorisation of intents. Integrated in a response template, it could be exploited to enable a robot to provide convenient information.

Table 2. Three defined levels of intents and their implicit intent, each illustrated on an exemplary utterance.

Level	Intent	Example
Potential	Capabilities	*What can you do?*
Process	Explain_process	*I would like to learn how you welcome people.*
Decision	Robot_experience	*and what did you do after you noticed that?*

5 Related Work

We subsequently discuss some important academic publications related to this multidisciplinary research, including human-robot interaction, robot explainability and Conversation Analysis (CA), in order to put it in a larger context for discussion.

As Langley (2016) argues, robots engaging in explainable agency do not have to do it using a human language, but communication must be managed in some form that is easy to understand for a human [16]. With regard to the locality of human-robot interaction, this research relates to the category of remote interaction interfaces [11], because there is no need for temporal or spatial co-location of robot and user. Pepper executes tasks automatically, informs users and has means to adapt its course of action. Thus, the level of its autonomy, which determines how interaction between robots and humans is established and designed [3], is quite high here. Even though the case study includes a social robot in public spaces, it rather contributes to perception and interaction methods in computer-mediated communication [10] than to social robotics (e.g. [32]).

Consequently, we state our work to contribute to approaches in AI and robotics to improve the *explainability of autonomous and complex technical systems using a remote Conversational Interface before and after their mission.*

There is already some remarkable research going on, paying attention to human-computer communication via Conversational Interfaces. Zhou et al. recently reported a WoZ field study where user perception and interaction was investigated in an apparently quite similar setting [33]. While in that case the chatbot (or the Wizard) was the interviewer and users were respondents, we looked at the participants as the information seeker. Also the focus of analysis was more on how the user perceives the chatbot's behaviour than on how s/he utters a request.

Explainability has a long tradition in AI and dates back to, for example, expert and case-based reasoning systems in the 80s and 90s described in [4,30]. These systems were able to make their conclusions about recommendations and decisions transparent. With the advent of AI-based systems such as autonomous cars and service robots there is resurgence in the field of explainable AI [18, 20]. However, as Miller points out in [21], a majority of approaches focuses on what an useful or good explanation is from the researchers perspective who, for example, developed an algorithm or method. The actual user is rarely taken into account, even though the existence of individual differences in demands is evident [15]. Consequently, researchers' requirements for a 'good' interface remain shallow. For example in [23], a learning-based approach is presented to answer questions about the task history of a robot, where questions were mainly driven by availability of data instead of users' needs. In the present investigation we chose an user-centred design perspective.

Conversation Analysis (CA) looks at language as interactional resource, and the interaction itself as sequentially organised social actions [26]. While CA has already been effectively used in human-robot interaction domains [24], its potential for the development of Conversational Interfaces remained widely unexploited up to now.

Usually, chatbot designers try to foresee all possible types of user questions by mapping them (directly or indirectly) to a set of utterance categories (called *intents*) that help to manage natural language understanding (NLU). More sophisticated technologies, such as dialogue management and semantic analysis, can be used to make the system 'smarter' [19]. However, this is usually connected to large linguistic resources, domain knowledge and very complex analysis that makes the system slow. As an alternative, [13] showed how computational models of dialogue can be created from a small number of examples using CA for the analysis: the author described turn formats as a set of abstract rules that can be filled with different sets of interaction devices and are, in this way, even language independent. We adopt a similar approach in this study.

The concept of *recipient design* helps to analyse the speakers' choices of interactional resources to make their utterances correctly understandable for the recipient [14]. This again is largely influenced by epistemic stances [12], which describe a speaker's expectation about what the other speaker may know. Applied to the present scenario, where a *machine* is on the other end of the line instead of a human, participants' utterances provide insights to their demands, beliefs and perceptions towards the chatbot.

6   Discussion

According to the hypotheses stated in Sect. 1, (1) different characteristics for the classification of requests could successfully be identified, as for instance the level of abstraction or the scope of validity (Sect. 3.1). Fundamentally different strategies in eliciting information were observed and described in Sect. 3.2. Furthermore, (2) associated requirements, risks and challenges were identified and substantiated with particular chat sequences in Sect. 4 and pave the road map for the development of a successfully interacting conversational agent.

First, there need to be a mechanism to *handle unresponded questions* (repeat, modify or forget). This might include any form of prediction, to enable the agent to factor sequential consequences into decision. Second, there is a need for an appropriate *recognition of intents*. Those are formulated by the human as direct or indirect requests depending on the sequential position. Finally, strategies for robot-initiated sequences to channel the conversation reasonably are required. This way, the robot can offer information and focus on what it *can* do, while the user may decide to accept the offer or to change direction.

The chosen method for experimental design carries both advantages and limitations. Even though most established statistical magnitudes for evaluation are unsuitable for such qualitative approaches, we can still discuss its internal and external qualitative characteristics.

It is possible to create valid models of dialogue even from a small number of examples using methods of CA. In this way, this study confirms the validity of the method introduced in [13]. All participants including the Wizard were non-native English speaker, which can be considered as both an advantage or a limitation. A native speaker might have a more acute sense for subtleties, however such a system needs to be generally applicable and robust against the individual user background. Although there were instructions and sample answers provided for the Wizard, a more detailed behavioural definition would be helpful, to enhance comparability and significance of results. These instructions would be very fine-grained and should ideally be provided in form of response templates and instructions related to turn-taking behaviour. Observations and conclusions of this case study are evidently transferable to other domains to a certain extent. Some aspects, as the defined types of intents, are highly context related and thus individual. Still, the overall concept of processing user requests can be generalised. Likewise, the sequential structure of interaction is independent of the system in the back end. Overcoming the identified obstacles can serve as a general step towards more intelligent Conversational Interfaces. Even in this comparably small dataset, we observed users not following the instructions. Consequently, even task-based Conversational Interfaces need to implement special policies to handle unexpected requests to become more robust and keep the conversation focused.

In contrast to the general tendency in NLP to use large corpora for modelling, the present study confirms that rule-based or hybrid systems can successfully be designed from very small corpora.

7   Conclusion and Outlook

In this article we present an exploratory Wizard of Oz study for human-robot interaction via Conversational Interfaces with the purpose to foster robot explainability. We focused on the user behaviour and applied Conversation Analysis to create a functional specification for such an interface from a small number of examples.

According to the nature of exploratory research, we identified important key aspects for both practical implementation and further well-founded investigations. We demonstrated successfully that users of an artificially intelligent system may formulate their request in several different ways. Even though their task is quite basic and clearly defined, humans tend to ask for the desired information implicitly, instead of formulating a straightforward question. Based on the discussed findings, we formulated features that are to be considered for the implementation of a Conversational Interface.

Participants showed remarkably strong interest in the release of the chatbot, which we pretended to test here. Thus, we feel confirmed in our belief that there is a need for such systems. We are currently working on the actual implementation of a Conversational Interface and experimenting with different frameworks and tools available on the market such as Watson, RASA and others. We aim to realise the identified findings and requirements.

References

1. Anjomshoae, S., Najjar, A., Calvaresi, D., Främling, K.: Explainable agents and robots: results from a systematic literature review. In: AAMAS 2019, July 2019
2. Austin, J.L.: How to Do Things with Words. Clarendon Press, Oxford (1962)
3. Beer, J.M., Fisk, A.D., Rogers, W.A.: Toward a framework for levels of robot autonomy in human-robot interaction. J. Hum.-Robot Interact. **3**(2), 74–99 (2014)
4. Chandrasekaran, B., Tanner, M.C., Josephson, J.R.: Explaining control strategies in problem solving. IEEE Expert: Intell. Syst. Appl. **4**(1), 9–15 (1989). 19–24
5. Di Prospero, A., Norouzi, N., Fokaefs, M., Litoiu, M.: Chatbots as assistants: an architectural framework. In: Proceedings of the 27th Annual International Conference on Computer Science and Software Engineering, CASCON 2017, pp. 76–86. IBM Corp., Riverton (2017)
6. Dingemanse, M., Blythe, J., Dirksmeyer, T.: Formats for other-initiation of repair across languages: an exercise in pragmatic typology. Stud. Lang. **3**(81), 5–43 (2014)
7. Došilović, F.K., Brčić, M., Hlupić, N.: Explainable artificial intelligence: a survey. In: 2018 41st International Convention on Information and Communication Technology, Electronics and Microelectronics (MIPRO), pp. 0210–0215. IEEE (2018)
8. Došilović, F.K., Brčić, M., Hlupić, N.: Explainable artificial intelligence: a survey. In: 2018 41st International Convention on Information and Communication Technology, Electronics and Microelectronics (MIPRO), pp. 0210–0215, May 2018. https://doi.org/10.23919/MIPRO.2018.8400040
9. Fernandez, A., Herrera, F., Cordon, O., del Jesus, M.J., Marcelloni, F.: Evolutionary fuzzy systems for explainable artificial intelligence: why, when, what for, and where to? IEEE Comput. Intell. Mag. **14**(1), 69–81 (2019)
10. González-Lloret, M.: Conversation analysis of computer-mediated communication. CALICO **28**(2), 308–325 (2011)
11. Goodrich, M.A., Schultz, A.C.: Human-robot interaction: a survey. Found. Trends Hum.-Comput. Interact. **1**(3), 203–275 (2007)
12. Heritage, J.: The epistemic engine: sequence organization and territories of knowledge. Res. Lang. Soc. Interact. **45**(1), 30–52 (2012)
13. Höhn, S.: Data-driven repair models for text chat with language learners. Ph.d. thesis, University of Luxembourg (2016)
14. Hutchby, I.: Aspects of recipient design in expert advice-giving on call-in radio. Discourse process. **19**(2), 219–238 (1995)
15. Kaptein, F., Broekens, J., Hindriks, K., Neerincx, M.: Personalised self-explanation by robots: the role of goals versus beliefs in robot-action explanation for children and adults. In: 2017 26th IEEE International Symposium on Robot and Human Interactive Communication (RO-MAN), pp. 676–682. IEEE (2017)
16. Langley, P.: Explainable agency in human-robot interaction. In: AAAI Fall Symposium Series. AAAI Press (2016)
17. Langley, P.: Explainable, normative, and justified agency (2019)
18. Langley, P., Meadows, B., Sridharan, M., Choi, D.: Explainable agency for intelligent autonomous systems. In: AAAI, pp. 4762–4764 (2017)
19. MacTear, M., Callejas, Z., Griol, D.: The Conversational Interface: Talking to Smart Devices. Springer, Heidelberg (2016). https://doi.org/10.1007/978-3-319-32967-3
20. Miller, T.: Explanation in artificial intelligence: insights from the social sciences. Artif. Intell. 267, 1–38 (2018), arXiv:1706.07269

21. Miller, T., Howe, P., Sonenberg, L.: Explainable AI: beware of inmates running the asylum. In: IJCAI 2017 Workshop on Explainable Artificial Intelligence (XAI) (2017), http://people.eng.unimelb.edu.au/tmiller/pubs/explanation-inmates.pdf
22. Neerincx, M.A., van der Waa, J., Kaptein, F., van Diggelen, J.: Using perceptual and cognitive explanations for ecnhanced human-agent team performance. In: Harris, D. (ed.) EPCE 2018. LNCS (LNAI), vol. 10906, pp. 204–214. Springer, Cham (2018). https://doi.org/10.1007/978-3-319-91122-9_18
23. Perera, V., Veloso, M.: Learning to understand questions on the task history of a service robot. In: Proceedings of RO-MAN 2017, the IEEE International Symposium on Robot and Human Interactive Communication, Lisbon, Portugal, August 2017
24. Pitsch, K., Wrede, S.: When a robot orients visitors to an exhibit. referential practices and interactional dynamics in real world HRI. In: The 23rd IEEE International Symposium on Robot and Human Interactive Communication, 2014 RO-MAN, pp. 36–42 (2014)
25. Riek, L.D.: Wizard of oz studies in HRI: a systematic review and new reporting guidelines. J. Hum.-Robot Interact. 1(1), 119–136 (2012)
26. Schegloff, E.A.: Sequence Organization in Interaction: A Primer in Conversation Analysis, Vol. 1, 1 edn. Cambridge University Press (2007)
27. Shahriari, K., Shahriari, M.: Ieee standard review – ethically aligned design: a vision for prioritizing human wellbeing with artificial intelligence and autonomous systems. In: 2017 IEEE Canada International Humanitarian Technology Conference (IHTC), pp. 197–201, July 2017. https://doi.org/10.1109/IHTC.2017.8058187
28. Steinfeld, A., Jenkins, O.C., Scassellati, B.: The oz of wizard: simulating the human for interaction research. In: Proceedings of the 4th ACM/IEEE International Conference on Human Robot Interaction, pp. 101–108. ACM (2009)
29. Stivers, T., Enfield, N.J.: A coding scheme for question-response sequences in conversation. J. Pragmatics 42(10), 2620–2626 (2010)
30. Swartout, W., Paris, C., Moore, J.: Explanations in knowledge systems: design for explainable expert systems. IEEE Expert 6(3), 58–64 (1991)
31. Tudini, V.: Online Second Language Acquisition: Conversation Analysis of Online Chat. Continuum (2010)
32. Yan, H., Ang, M.H., Poo, A.N.: A survey on perception methods for human-robot interaction in social robots. Int. J. Soc. Robot. 6(1), 85–119 (2014)
33. Zhou, M.X., Wang, C., Mark, G., Yang, H., Xu, K.: Building real-world chatbot interviewers: Lessons from a wizard-of-oz field study (2019)

Opening the Black Box

Explanations of Black-Box Model Predictions by Contextual Importance and Utility

Sule Anjomshoae[(✉)], Kary Främling, and Amro Najjar

Department of Computing Science, Umeå University, Umeå, Sweden
{sule.anjomshoae,kary.framling,amro.najjar}@umu.se

Abstract. The significant advances in autonomous systems together with an immensely wider application domain have increased the need for trustable intelligent systems. Explainable artificial intelligence is gaining considerable attention among researchers and developers to address this requirement. Although there is an increasing number of works on interpretable and transparent machine learning algorithms, they are mostly intended for the technical users. Explanations for the end-user have been neglected in many usable and practical applications. In this work, we present the Contextual Importance (CI) and Contextual Utility (CU) concepts to extract explanations that are easily understandable by experts as well as novice users. This method explains the prediction results without transforming the model into an interpretable one. We present an example of providing explanations for linear and non-linear models to demonstrate the generalizability of the method. CI and CU are numerical values that can be represented to the user in visuals and natural language form to justify actions and explain reasoning for individual instances, situations, and contexts. We show the utility of explanations in car selection example and Iris flower classification by presenting complete (i.e. the causes of an individual prediction) and contrastive explanation (i.e. contrasting instance against the instance of interest). The experimental results show the feasibility and validity of the provided explanation methods.

Keywords: Explainable AI · Black-box models · Contextual importance · Contextual utility · Contrastive explanations

1 Introduction

Intelligent systems are widely used for decision support across a broad range of industrial systems and service domains. A central issue that compromises the adoption of intelligent systems is the lack of explanations for the actions taken by them. This is a growing concern for effective human-system interaction. Explanations are particularly essential for intelligent systems in medical diagnosis, safety-critical industry, and automotive applications as it raises trust and transparency in the system. Explanations also help users to evaluate the accuracy of the system's predictions [1]. Due to a growing need for intelligent systems' explanations, the field of eXplainable Artificial Intelligence (XAI) is receiving a considerable amount of attention among developers and researchers [2].

© Springer Nature Switzerland AG 2019
D. Calvaresi et al. (Eds.): EXTRAAMAS 2019, LNAI 11763, pp. 95–109, 2019.
https://doi.org/10.1007/978-3-030-30391-4_6

While generating explanations have been investigated in early years of expert systems, intelligent systems today have become immensely complex and rapidly evolving in new application areas. As a result, generating explanation for such systems is more challenging and intriguing than ever before [3, 4]. This is particularly relevant and important in intelligent systems that have more autonomy in decision making. Nonetheless, as important as it is, existing works are mainly focusing on either creating mathematically interpretable models or converting black-box algorithms into simpler models. In general, these explanations are suitable for expert users to evaluate the correctness of a model and are often hard to interpret by novice users [5, 6]. There is a need for systematic methods that considers the end user requirements in generating explanations.

In this work, we present the Contextual Importance (CI) and Contextual Utility (CU) methods which explain prediction results in a way that both expert and novice users can understand. The CI and CU are numerical values which can be represented as visuals and natural language form to present explanations for individual instances [7]. Several studies suggested modeling explanation facilities based on practically relevant theoretical concepts such as contrastive justifications to produce human understandable explanations along with the complete explanations [8]. Complete explanations present the list of causes of an individual prediction, while contrastive explanations justify why a certain prediction was made instead of another [9]. In this paper, we aim at providing complete explanations as well as the contrastive explanations using CI and CU methods for black-box models. This approach generally can be used with both linear and non-linear learning models. We demonstrate an example of car selection problem (e.g. linear regression) and classification problem (e.g. neural network) to extract explanations for individual instances.

The rest of the paper is organized as follows: Sect. 2 discusses the relevant background study. Section 3 reviews the state of the art for generating explanation. Section 4 explains the contextual importance and utility method. Section 5 presents the explanation results for the regression and classification example. Section 6 discusses the results and, Sect. 7 concludes the paper.

2 Background

Explanations were initially discussed in rule-based expert systems to support developers for system debugging. Shortliffe's work is probably the first to provide explanation in a medical expert system [10]. Since then, providing explanations for intelligent systems' decisions and actions has been a concern for researchers, developers and the users. Earlier attempts were limited to traces, and a line of reasoning explanations that are used by the decision support system. However, this type of explanations could only be applied in rule-based systems and required knowledge of decision design [11]. These systems were also unable to justify the rationale behind a decision.

Swartout's framework was one of the first study that emphasized the significance of justifications along with explanations [12]. Early examples proposed justifying the outcomes through drilling-down into the rationale behind each step taken by the system. One approach to produce such explanation was storing the justifications as canned text for all the possible questions that can be inquired [13]. However, this approach had several drawbacks such as maintaining the consistency between the model and the explanations, and predicting all the possible questions that the system might encounter.

The decision theory was proposed to provide justifications for the system's decision. Langlotz suggested decision trees to capture uncertainties and balance between different variables [14]. Klein developed explanation strategies to justify value-based preferences in the context of intelligent systems [15]. However, these explanations required knowledge of the domain in which the system will be used [16]. This kind of explanation were less commonly used, due to the difficulties in generating such explanations that satisfies the needs of the end-users [11].

Expert systems that are built based on probabilistic decision-making systems such as Bayesian networks required the explanations even more due to their internal logic is unpredictable [17]. Comprehensive explanations of probabilistic reasoning are therefore studied in a variety of applications to increase the acceptance of expert systems [18]. Explanation methods in Bayesian networks have been inadequate to constitute a standard method which is suitable for systems with similar reasoning techniques.

Previous explanation studies within expert systems are mostly based on strategies that rely on knowledge base and rule extraction. However, these rule-based systems and other symbolic methods perform poorly in many explanation tasks. The number of rules tends to grow extremely high, while the explanations produced are limited to showing the applicable rules for the current input values. The Contextual Importance and Utility (CIU) method was proposed to address these explanation problems earlier [7]. This method explains the results directly without transforming the knowledge. The details of this work are discussed in Sect. 4.

3 State of the Art

Machine learning algorithms are the heart of many intelligent decision support systems in finance, medical diagnosis, and manufacturing domains. Because some of these systems are considered as black-box (i.e. hiding inner-workings), researchers have been focusing on integrating explanation facilities to enhance the utility of these systems [5]. Recent works define interpretability particular to their explanation problems. Generally these methods are categorized into two broad subject-matter namely, model-specific and model-agnostic methods. The former one typically refers to inherently interpretable models which provide a solution for a predefined problem. The latter provides generic framework for interpretability which is adaptable to different models.

In general, model-specific methods are limited to certain learning models. Some intrinsically interpretable models are sparse linear models [19, 20], discretization methods such as decision trees and association rule lists [21, 22], and Bayesian rule lists [23]. Other approaches include instance-based models [24] and mind-the-gap model [25] focus on creating sparse models through feature selection to optimize interpretability. Nevertheless, linear models are not that competent at predictive tasks, because the relationships that can be learned are constrained and the complexity of the problem is overgeneralized. Even though they provide insight into why certain predictions are made, they enforce restrictions on the model, features, and the expertise of the users.

Several model-agnostic frameworks have been recently proposed as an alternative to interpretable models. Some methods suggest measuring the effect of an individual feature on a prediction result by perturbing inputs and seeing how the result changes [26, 27]. The effects are then visualized to explain the main contributors for a prediction and to compare the effect of the feature in different models. Ribeiro et al. [28] introduce Locally Interpretable Model Explanation (LIME) which aims to explain an instance by approximating it locally with an interpretable model. The LIME method implements this by sampling around the instance of interest until they arrive at a linear approximation of the global decision function. The main disadvantage of this method is that data points are sampled without considering the correlation between features. This can create irrelevant data points which can lead to false explanations. An alternative method is Shapley values where the prediction is fairly distributed among the features based on how each feature contributes to the prediction value. Although, this method generates complete and contrastive explanations, it is computationally expensive. In general, model-agnostic methods are more flexible than model-specific ones. Nevertheless, the correctability of the explanations and incorporating user feed-back in explanation system are still open research issues [29].

4   Contextual Importance and Contextual Utility

Contextual importance and utility were proposed as an approach for justifying recommendations made by black-box systems in Kary Främling's PhD thesis [30], which is presumably one of the earliest studies addressing the need to explain and justify specific recommendations or actions to end users. The method was proposed to explain preferences learned by neural networks in a multiple criteria decision making (MCDM) context [31]. The real-world decision-making case consisted in choosing a waste disposal site in the region of Rhône-Alpes, France, with 15 selection criteria and over 3000 potential sites to evaluate. A similar use case was implemented based on data available from Switzerland, as well as a car selection use case. In such use cases, it is crucial to be able to justify the recommendations of the decision support system also in ways that are understandable for the end-users, in this case including the inhabitants of the selected site(s).

Multiple approaches were used for building a suitable MCDM system, i.e. the well-known MCDM methods Analytic Hierarchy Process (AHP) [32] and ELECTRE [33]. A rule-based expert system was also developed. However, all these approaches suffer from the necessity to specify the parameters or rules of the different models, which needs to be based on a consensus between numerous experts, politicians and other stakeholders. Since such MCDM systems can always be criticized for being subjective, a machine learning approach that would learn the MCDM model in an "objective" way based on data from existing sites became interesting.

MCDM methods such as AHP are based on *weights* that express the *importance* of each input (the selection criteria) for the final decision. A notion of *utility* and *utility function* is used for expressing to what extent different values of the selection criteria are favorable (or not) for the decision. Such MCDM methods are linear in nature, which limits their mathematical expressiveness compared to neural networks, for instance. On the other hand, the weights and utilities give a certain transparency, or explainability, to the results of the system. The rationale behind Contextual Importance (CI) and Contextual Utility (CU) is to generalize these notions from linear models to non-linear models [7].

In practice, the importance of criteria and the usefulness of their values change according to the current context. In cold weather, the importance and utility of warm clothes increases compared to warm summer weather, whereas the importance of the sunscreen rating that might be used becomes small. This is the reason for choosing the word contextual to describe CI and CU. This approach generally can be used with both linear and non-linear learning models. It is based on explaining the model's predictions on individual importance and utility of each feature.

CI and CU are defined as:

$$CI = \frac{Cmax_x(C_i) - Cmin_x(C_i)}{absmax - absmin} \tag{1}$$

$$CU = \frac{y_{i,j} - Cmin_x(C_i)}{Cmax_x(C_i) - Cmin_x(C_i)} \tag{2}$$

where

- C_i is the context studied (which defines the fixed input values of the model),
- x is the input(s) for which CI and CU are calculated, so it may also be a vector,
- $y_{i,j}$ is the output value for the output j studied when the inputs are those defined by C_i,
- $Cmax_x(C_i)$ and $Cmin_x(C_i)$ are the highest and the lowest output values observed by varying the value of the input(s) x,
- $absmax$ and $absmin$ specify the value range for the output j being studied.

CI corresponds to the fraction of output range covered by varying the value(s) of inputs x and the maximal output range. CU reflects the position of $y_{i,j}$ within the output range covered ($Cmax_x(C_i) - Cmin_x(C_i)$). Each feature x with prediction $y_{i,j}$ has its own CI and CU values.

The estimation of $Cmax_x(C_i)$ and $Cmin_x(C_i)$ is a mathematical challenge, which can be approached in various ways. In this paper, we have used Monte-Carlo simulation, i.e. generating a "sufficient" number of input vectors with random values for the x input(s). Obtaining completely accurate values for $Cmax_x(C_i)$ and $Cmin_x(C_i)$ would in principle require an infinite number of random values. However, for the needs of explainability, it is more relevant to obtain CI values that indicate the relative importance of inputs compared to each other. Regarding CU, it is not essential to obtain exact values neither for producing appropriate explanations. However, the estimation of $Cmax_x(C_i)$ and $Cmin_x(C_i)$ remains a matter of future studies. Gradient-based methods might be appropriate in order to keep the method model-agnostic. In [30], Normalized Radial Basis Function (RBF) networks were used, where it makes sense to assume that minimal and maximal output values will be produced at or close to the centroids of the RBF units. However, such methods are model-specific, i.e. specific to a certain type or family of black-box models.

CI and CU are numerical values that can be represented in both visual and textual form to present explanations for individual instances. CI and CU can also be calculated for more than one input or even for all inputs, which means that arbitrary higher-level concepts that are combinations of more than one inputs can be used in explanations. Since the concepts and vocabularies that are used for producing explanations are external to the black box, the vocabularies and visual explanations can be adapted depending on the user they are intended for. It is even possible to change the representation used in the explanations if it turns out that the currently used representation is not suitable for the user's understanding, which is what humans tend to do when another person does not seem to understand already tested explanation approaches. Figure 1 illustrates how explanations are generated using contextual importance and utility method.

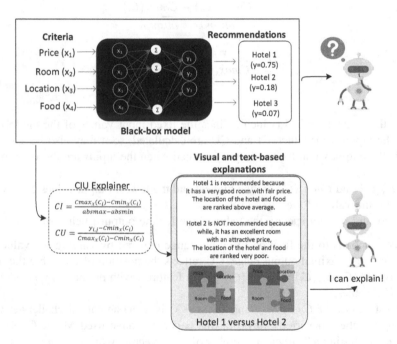

Fig. 1. Providing explanations for individual instances using CI and CU

Another important point is that humans usually ask for explanations of why a certain prediction was made instead of another. This gives more insight into what would be the case if the input had been different. Creating contrastive explanations and comparing the differences to another instance can often be more useful than the complete explanation alone for a particular prediction. Since the contextual importance and utility values can be produced for all possible input value combinations and outputs, it makes it possible to explain why a certain instance C_i is preferable to another one, or why one class (output) is more probable than another. The algorithm used for producing complete and contrastive explanations is shown in Fig. 2.

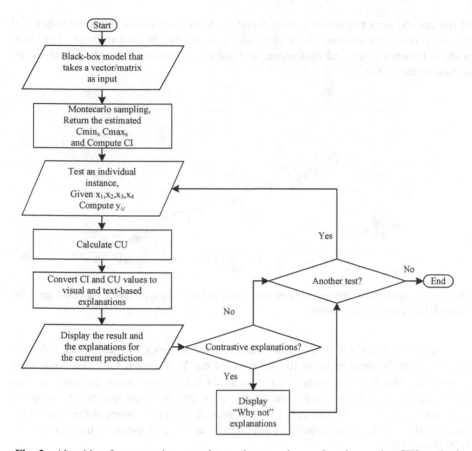

Fig. 2. Algorithm for generating complete and contrastive explanations using CIU method

5   Examples of CI and CU Method to Extract Explanation for Linear and Non-linear Models

The explanation method presented here provides flexibility to explain any learning model that can be considered a "black-box". In this section, we present the examples of providing explanations for linear and non-linear models using contextual importance and utility method. Code explaining individual prediction for non-linear models is available at https://github.com/shulemsi/CIU.

5.1   Visual Explanations for Car Selection Using CI and CU Method

The result of explanations for a car selection problem using CI and CU method is presented. The dataset used in this example was initially created and utilized to learn the preference function by neural network in a multi-criteria decision-making problem [30]. Here, these samples are used to show how explanations can be generated for linear models. The dataset contains 113 samples with thirteen different characteristics of the car and their respective scores. Some of the characteristics are namely; price of the car, power, acceleration, speed, dimensions, chest, weight, and aesthetic. The linear relation between price and preference, and the corresponding CI and CU values are demonstrated in Fig. 3.

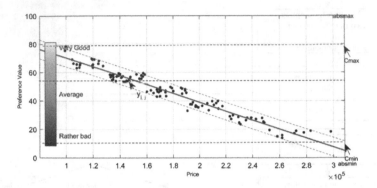

Fig. 3. The contextual importance and contextual utility values for price (selected car WW Passat GL). (Color figure online)

The preference value is shown as a function of the price of the car, the red-cross ($y_{i,j}$) showing the current value for the selected car WW Passat GL. The color scale shows the limits for translating contextual utility values into words to generate text-based explanations. Contextual importance values are converted into words using the same kind of scale. Table 1 reveals $Cmax_x$, $Cmin_x$, CI and CU values of the price of the car and other key features including power, acceleration, and speed for the example car WW Passat GL.

Table 1. CI and CU values of the features price, power, acceleration, and speed for the selected car example WW Passat-GL

	Price	Power	Acceleration	Speed
Cmin	13	14	13	10
Cmax	79	78	68	64
CI%	66	64	55	54
CU	0.67	0.15	0.30	0.25

The table shows that the price and power are the most important features for the selected car. Also, the highest utility value belongs to the price which means it is the most preferred feature of this car. The least preferred characteristic of this car is the power which has the lowest utility value. The CI and CU values led the following visual and text-based explanations as shown in Fig. 4.

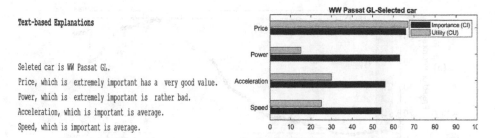

Fig. 4. Text-based and visual explanations for selected car WW Passat GL

The contrastive explanations are generated to compare the selected car example to other instances. Comparison between the selected car (WW Passat GL) to the expensive car (Citroen XM), and to the average car (WW Vento) is visually presented. Figure 5 shows that the selected car has a very good value compare to the average car considering the importance of this criteria. Although the average car has higher utility values for acceleration and speed, it is exceeding the importance of the criteria. Similarly, the expensive car has very low utility in terms of price, and it has quite high values for power, acceleration and speed compare to the selected car.

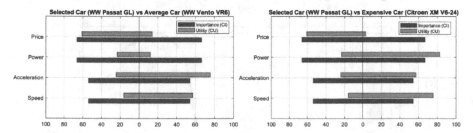

Fig. 5. Contrastive visual explanations for selected car (WW Passat GL) with average car (WW Vento VR6) and expensive car (Citroen XM V6-24)

5.2 Explaining Iris Flower Classification Using CI and CU Method

In this section, the results of explaining individual predictions using CI and CU on Iris flower classification is presented. The dataset contains 150 labeled flowers from the genus Iris. The trained network classifies Iris flowers into three species; Iris Setosa, Iris Versicolor and Iris Virginica based on the properties of leaves. These properties are namely; petal length, petal width, sepal length, sepal width. The trained network outputs the prediction value for each species which the highest one being the predicted class. The network is used to classify patterns that it has not seen before and results are used to generate explanations for individual instances.

An example of how explanations are generated based on CI and CU values is illustrated in Fig. 6. Given following input values; 7 (petal length), 3.2 (petal width), 6 (sepal length), 1.8 (sepal width), model predicts the class label as Iris Virginica. In order to compute the CI and CU values, we randomize 150 samples, and estimate $Cmax_x(C_i)$ and $Cmin_x(C_i)$ values for each input feature. The red-cross $(y_{i,j})$ indicates

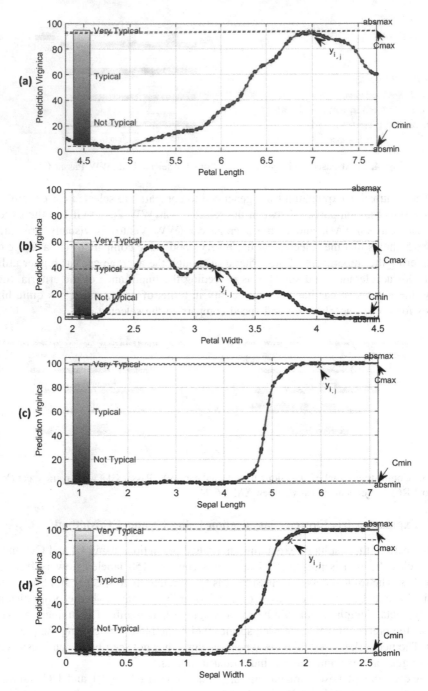

Fig. 6. CI and CU values of each features for Iris Virginica classification, (a) Petal length, (b) Petal width, (c) Sepal length, and (d) Sepal width (Color figure online)

the current prediction value. Each figure demonstrates the importance of that feature and the usefulness for the predicted class. Similarly, CI and CU values of other classes are obtained to generate contrastive explanations. Note that a feature that is distinguishing for Iris Virginica may not be that distinguishing or important for other classes. The color bar indicates the contextual utility values converted into natural language expressions.

Table 2 shows the result of the sample test. For this case, sepal length is the most important feature with the highest utility value contributing to the class and the petal width is the least contributing feature for the given instance.

Table 2. $Cmin_x$, $Cmax_x$ and CIU values of each feature for the class label Iris Virginica

	Petal length	Petal width	Sepal length	Sepal width
Cmin	3	1	0	0
Cmax	92	56	100	100
CI%	89	55	100	100
CU	1 ($y_{i,j} = .92$)	0.69 ($y_{i,j} = .39$)	1 ($y_{i,j} = 1$)	0.91 ($y_{i,j} = .91$)

Table 3 shows how these values are transformed into natural language expressions to generate explanations based on the degree of the values.

Table 3. Symbolic representation of the CI and CU values

Degree (d)	Contextual importance	Contextual utility
$0 < d \leq 0.25$	Not important	Not typical
$0.25 < d \leq 0.5$	Important	Unlikely
$0.5 < d \leq 0.75$	Rather important	Typical
$0.75 < d \leq 1.0$	Highly important	Very typical

The obtained values are translated into explanation phrases as shown in Fig. 7. These are the complete explanations which justifies why the model predicts this class label. Furthermore, the contrastive explanations are produced to demonstrate the contrasting cases. Figure 8 shows the results of this application.

```
The model`s prediction is 98% Iris Virginica. Because;
The petal length  which is a highly important (CI=89%) feature has a very typical (CU=1) size.
The petal width  which is rather an important (CI=55%) feature has a typical (CU=0.69) size.
The sepal length  which is a highly important (CI=100%) feature has a very typical (CU=1) size.
The sepal width  which is a highly important (CI=100%) feature has a very typical (CU=0.91) size.
And the biggest contributing feature is the sepal length.
```

Fig. 7. Complete explanation for the class label Iris Virginica

```
It is not Iris Setosa(0%), because;
The petal length  which is a highly important (CI=86%) feature has not a typical (CU=0) size.
The petal width  which is a highly important (CI=98%) feature has an unlikely (CU=0.48) size.
The sepal length  which is a highly important (CI=100%) feature has not a typical (CU=0) size.
The sepal width  which is a highly important (CI=100%) feature has not a typical (CU=0) size.
It is not Iris Versicolor(2%), because;
The petal length  which is rather an important (CI=61%) feature has not a typical (CU=0.03) size.
The petal width  which is a highly important (CI=97%) feature has not a typical (CU=0.13) size.
The sepal length  which is a highly important (CI=99%) feature has not a typical (CU=0) size.
The sepal width  which is a highly important (CI=100%) feature has not a typical (CU=0.09) size.
```

Fig. 8. Contrastive explanations for Iris Setosa and Iris Versicolor

6   Discussion

Intelligent systems that are explaining their decisions to increase the user's trust and acceptance are widely studied. These studies propose various means to deliver explanations in form of; if-then rules [34], heat-maps [35], visuals [36], and human-labeled text [37]. These explanations and justifications provide limited representation of the cause of a decision. The CIU method presented here proposes two modalities as visuals and textual form to express relevant explanations. The variability in modality of presenting explanations could improve interaction quality, particularly in time-sensitive situations (e.g. switching to visual explanations from text-based explanations). More-over, CI and CU values can be represented with different levels of details and produce explanations that are tailored to the users' specification. User-customized explanations could reduce ambiguity in reasoning. This is particularly important in safety-critical applications where users require a clear response from the system.

Explanation methods should be responsive to different types of queries. Most explanation methods only provide explanations which respond to why a certain deci-sion or prediction was made. However, humans usually expect explanations with a contrasting case to place the explanation into a relevant context [8]. This study present examples of complete and contrastive explanation to justify the predicted outcomes. One stream of research propose justification based explanations for image dataset combining visual and textual information [38]. Although they produce convincing explanations for users, they offer post-hoc explanation which is generally constructed without following the model's reasoning path (unfaithfully).

Faithfulness to actual model is important to shows the agreement to the input-output mapping of the model. If the explanation method is not faithful to the original model then the validity of explanations might be questionable. While the rule extrac-tion method produces faithful explanations, it is often hard to trace back the reasoning path, particularly when the number of features is too high. Other methods such as approximating an interpretable model provide only local fidelity for individual instances [28]. However, features that are locally important may not be important in the global context. CIU overcome the limitation of the above methods by providing explanations based on the highest and the lowest output values observed by varying the value of the input(s). However, accurate estimation of the minimal and the maximal values remains a matter of future studies. Furthermore, CIU is a model agnostic method which increases the generalizability of the explanation method in selection of the learning model.

7 Conclusion

The aim of this paper is presenting contextual importance and utility method to provide explanations for black-box model predictions. CIU values are represented as visuals and natural language expressions to increase the comprehensibility of the explanations. These values are computed for each class and features which enable to further produce contrastive explanation against the predicted class. We show the utilization of the CIU for linear and non-linear models to validate the generalizability of the method. Future work could extend the individual instance explanations to global model explanations in order to assess and select between alternative models. It is also valuable to focus on integrating CIU method into practical applications such as image labeling, recommender systems, and medical decision support systems. A future extension of our work relates to the CIU's utility in producing dynamic explanations by considering user's characteristics and investigating the usability of the explanations in real-world settings.

Acknowledgment. This work was partially supported by the Wallenberg AI, Autonomous Systems and Software Program (WASP) funded by the Knut and Alice Wallenberg Foundation.

References

1. Biran, O., Cotton, C.: Explanation and justification in machine learning: a survey. In: IJCAI-17 Workshop on Explainable AI (XAI) (2017)
2. Anjomshoae, S., Najjar, A., Calvaresi, D., Främling, K.: Explainable agents and robots: results from a systematic literature review. In: Proceedings of the 18th International Conference on Autonomous Agents and Multiagent Systems (2019)
3. Samek, W., Wiegand, T., Müller, K.R.: Explainable artificial intelligence: understanding, visualizing and interpreting deep learning models. arXiv preprint arXiv:1708.08296 (2017)
4. Ras, G., van Gerven, M., Haselager, P.: Explanation methods in deep learning: users, values, concerns and challenges. In: Escalante, H.J., et al. (eds.) Explainable and Interpretable Models in Computer Vision and Machine Learning. TSSCML, pp. 19–36. Springer, Cham (2018). https://doi.org/10.1007/978-3-319-98131-4_2
5. Nunes, I., Jannach, D.: Interaction, a systematic review and taxonomy of explanations in decision support and recommender systems. User Model. User-Adap. Interact. **27**(3–5), 393–444 (2017). https://doi.org/10.1007/s11257-017-9195-0
6. Miller, T., Howe, P., Sonenberg, L.: Explainable AI: beware of inmates running the asylum or: how I learnt to stop worrying and love the social and behavioural sciences. arXiv preprint arXiv:1712.00547 (2017)
7. Främling, K.: Explaining results of neural networks by contextual importance and utility. In: Proceedings of the AISB 1996 Conference (1996)
8. Miller, T.: Explanation in artificial intelligence: insights from the social sciences. In: Artificial Intelligence (2018)
9. Molnar, C.J.: Interpretable Machine Learning. A Guide For Making Black Box Models Explainable. Leanpub (2018)
10. Shortliffe, E.: Computer-Based Medical Consultations: MYCIN, vol. 2. Elsevier (2012)
11. Clancey, W.J.: The epistemology of a rule-based expert system—a framework for explanation. Artif. Intell. **20**(3), 215–251 (1983)

12. Swartout, W.R.: XPLAIN: a system for creating and explaining expert consulting programs. In: University of Southern California Marina Del Rey Information Sciences Institue (1983)
13. Swartout, W., Paris, C., Moore, J.: Explanations in knowledge systems: design for explainable expert systems. IEEE Expert **6**(3), 58–64 (1991)
14. Langlotz, C., Shortliffe, E.H., Fagan, L.M.: Using decision theory to justify heuristics. In: AAAI (1986)
15. Klein, D.A.: Decision-Analytic Intelligent Systems: Automated Explanation and Knowledge Acquisition. Routledge, Abingdon (2013)
16. Forsythe, D.E.: Using ethnography in the design of an explanation system. Expert Syst. Appl. **8**(4), 403–417 (1995)
17. Henrion, M., Druzdzel, M.J.: Qualtitative propagation and scenario-based scheme for exploiting probabilistic reasoning. In: Proceedings of the Sixth Annual Conference on Uncertainty in Artificial Intelligence. Elsevier Science Inc. (1990)
18. Lacave, C., Díez, F.J.: A review of explanation methods for Bayesian networks. Knowl. Eng. Rev. **17**(2), 107–127 (2002)
19. Souillard-Mandar, W., Davis, R., Rudin, C., Au, R., Penney, D.: Interpretable machine learning models for the digital clock drawing test. arXiv preprint arXiv:1606.07163 (2016)
20. Ustun, B., Rudin, C.: Supersparse linear integer models for optimized medical scoring systems. Mach. Learn. **102**(3), 349–391 (2016)
21. Lakkaraju, H., Bach, S.H., Leskovec, J.: Interpretable decision sets: a joint framework for description and prediction. In: Proceedings of the 22nd ACM SIGKDD International Conference on Knowledge Discovery and Data Mining (2016)
22. Rudin, C., Letham, B., Madigan, D.: Learning theory analysis for association rules and sequential event prediction. J. Mach. Learn. Res. **14**(1), 3441–3492 (2013)
23. Letham, B., Rudin, C., McCormick, T.H., Madigan, D.: Building interpretable classifiers with rules using Bayesian analysis. In: Department of Statistics Technical Report tr609, University of Washington (2012)
24. Kim, B., Rudin, C., Shah, J.A.: The Bayesian case model: a generative approach for case-based reasoning and prototype classification. In: Advances in Neural Information Processing Systems (2014)
25. Kim, B., Shah, J.A., Doshi-Velez, F.: Mind the gap: a generative approach to interpretable feature selection and extraction. In: Advances in Neural Information Processing Systems (2015)
26. Kononenko, I.: An efficient explanation of individual classifications using game theory. J. Mach. Learn. Res. **11**, 1–18 (2010)
27. Krause, J., Perer, A., Ng, K.: Interacting with predictions: visual inspection of black-box machine learning models. In: Proceedings of the 2016 CHI Conference on Human Factors in Computing Systems (2016)
28. Ribeiro, M.T., Singh, S., Guestrin, C.: Why should I trust you? In: Proceedings of the 22nd ACM SIGKDD International Conference on Knowledge Discovery and Data Mining - KDD 2016. pp. 1135–1144 (2016)
29. Ribeiro, M.T., Singh, S., Guestrin, C.: Model-agnostic interpretability of machine learning. arXiv preprint arXiv:1606.05386 (2016)
30. Främling, K.: Modélisation et apprentissage des préférences par réseaux de neurones pour l'aide à la décision multicritère. In: Institut National de Sciences Appliquées de Lyon, Ecole Nationale Supérieure des Mines de Saint-Etienne, France, p. 209 (1996)
31. Främling, K., Graillot, D.: Extracting explanations from neural networks. In: Proceedings of the ICANN (1995)
32. Saaty, T.L.: Decision Making for Leaders: The Analytic Hierarchy Process for Decisions in a Complex World. RWS Publications, Pittsburgh (1990)

33. Roy, B.: Classement et choix en présence de points de vue multiples. **2**(8), 57–75 (1968)
34. Setiono, R., Azcarraga, A., Hayashi, Y.: MofN rule extraction from neural networks trained with augmented discretized input. In: Proceedings of the 2014 International Joint Conference on Neural Networks (IJCNN), pp. 1079–1086 (2014)
35. Seo, D., Oh, K., Oh, I.S.: Regional multi-scale approach for visually pleasing explanations of deep neural networks. arXiv preprint arXiv:1807.11720 (2018)
36. Štrumbelj, E., Kononenko, I.: A general method for visualizing and explaining black-box regression models. In: Dobnikar, A., Lotrič, U., Šter, B. (eds.) ICANNGA 2011. LNCS, vol. 6594, pp. 21–30. Springer, Heidelberg (2011). https://doi.org/10.1007/978-3-642-20267-4_3
37. Berg, T., Belhumeur, P.N.: How do you tell a blackbird from a crow? In: IEEE International Conference on Computer Vision, pp. 9–16 (2013)
38. Huk Park, D., et al.: Multimodal explanations: justifying decisions and pointing to the evidence. In: Proceedings of the IEEE Conference on Computer Vision and Pattern Recognition (2018)

Explainable Artificial Intelligence Based Heat Recycler Fault Detection in Air Handling Unit

Manik Madhikermi, Avleen Kaur Malhi$^{(\boxtimes)}$, and Kary Främling

Aalto University, Helsinki, Finland
{manik.madhikermi,avleen.malhi,kary.framling}@aalto.fi

Abstract. We are entering a new age of AI applications where machine learning is the core technology but machine learning models are generally non-intuitive, opaque and usually complicated for people to understand. The current AI applications inability to explain is decisions and actions to end users have limited its effectiveness. The explainable AI will enable the users to understand, accordingly trust and effectively manage the decisions made by machine learning models. The heat recycler's fault detection in Air Handling Unit (AHU) has been explained with explainable artificial intelligence since the fault detection is particularly burdensome because the reason for its failure is mostly unknown and unique. The key requirement of such systems is the early diagnosis of such faults for its economic and functional efficiency. The machine learning models, Support Vector Machine and Neural Networks have been used for the diagnosis of the fault and explainable artificial intelligence has been used to explain the models' behaviour.

Keywords: Explainable artificial intelligence · Heat recycler unit · Support vector machine · Neural networks

1 Introduction

Heating, Ventilation and Air Conditioning (HVAC) systems count for 50% of the consumed energy in commercial buildings for maintaining indoor comfort [21]. Nonetheless, almost 15% of energy being utilized in buildings get wasted due to various faults (like control faults, sensor faults) which significantly occurs in HVAC systems because of lack of proper maintenance [20,28]. In HVAC systems, an Air Handling Unit (AHU) acts as a key component. The faults in heat recycler in AHU often go unnoticed for longer periods of time till the performance deteriorates which triggers the complaints related to comfort or equipment failure. There are various fault detection and diagnosis techniques being identified to benefit the owners which can reduce energy consumption, improve maintenance and increases effective utilization of energy. Heat recycler's faults can be detected by comparing the normal working condition data with the abnormal

© Springer Nature Switzerland AG 2019
D. Calvaresi et al. (Eds.): EXTRAAMAS 2019, LNAI 11763, pp. 110–125, 2019.
https://doi.org/10.1007/978-3-030-30391-4_7

data measured during heat recycler failure. Most of the fault detection methods have the training dataset as historical data for building machine learning models, such as support vector machine, neural networks, support vector machine or decision trees, depending on the training dataset. Abnormal data of heat recycler failure is identified as a different class from the normal working class by using various classification algorithms [9,27].

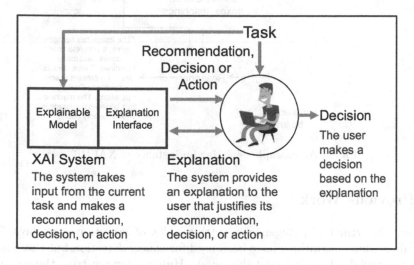

Fig. 1. The explanation framework of XAI [16]

We have entered a new era of artificial intelligence where core technology is machine learning but machine learning models are non-intuitive, opaque and it is difficult to understand them. Thus, the effectiveness of machine learning models is limited by its inability to give explanation for its behaviour. To overcome this, it is important for machine learning models to provide a human understandable explanation for explaining the rationale of model. This explanation can then further be used by analysts to evaluate if the decision meets the required rational reasoning and does not have reasoning conflicting with legal norms. But what does it mean by explanation?; a reason or justification given for some action. The explanation framework can be well explained by a framework as shown in Fig. 1 where XAI system consists of two modules, explanation model and explanation interface. The explanation model takes the input and justifies recommendation, decision or action based on any machine learning model. The explanation interface provides an explanation to justify the decision made by machine learning model i.e. why does the machine behaved in such particular way that made it to reach a particular decision. Thus the user can make the decision based on the explanation provided by the interface. Figure 2 shows an example that how explainable artificial intelligence helps a user by explaining the decisions of learning model.

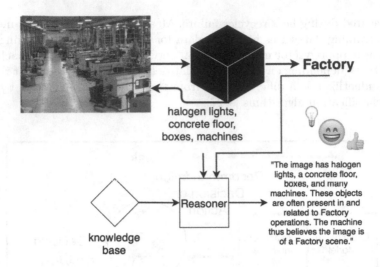

Fig. 2. An example depicting an instance of XAI [16]

2   Previous Work

Explainable artificial intelligence is getting a lot of attention nowadays. The machine learning algorithms have been used for cancer detection but these models do not explain the assessment they made. Humans can not trust these models since they do not understand the reason of their assessment [17]. Van Lent et al. [25] used the explanation capability in the training system developed by academic researchers and commercial game developers for the full spectrum command. Sneh et al. [24] used the explainable artificial intelligence in intelligent robotic systems for categorization of different types of errors. The errors have been divided into five categories using the machine learning techniques but they fail to provide the explanations. The XAI have been used to provide information and explanation of occurrence of these errors for three different machine learning models. Ten Zeldam et al. [31] proposed a technique for detection of incorrect or incomplete repair card in aviation maintenance that can result in failures. They proposed a Failure Diagnosis Explainability (FDE) technique for providing the interpretability and transparency to the learning model for the failure diagnosis. It is used to check if the accessed diagnosis can explain if a new failure detected matches the expected output of that particular diagnosis and if it is dissimilar to it, then it is not likely to be the real diagnosis.

A number of fault detection tools have currently emerged from research. Generally, stand-alone software product form is taken by these tools where there should be either offline processing of trend data or an online analysis can be provided for the building control system. There have been different data driven methods developed and used for detection of AHU's faults such as coil fouling, control valve fault, sensor bias etc. Yan et al. [30] proposed an unsupervised method for detecting faults in AHU by using cluster analysis. Firstly, PCA is used to reduce dimensions of collected historical data and then spacial separated data groups (clusters representing faults) are identified by using clustering algorithm. The proposed system was tested on a simulated data and was able to detect single and multiple faults in AHU.

Lee et al. [18] detected the AHU cooling coil subsystem's fault with the help of Artificial Neural Network (ANN) backward propagation method based on dominant residual signature. Wang et al. [26] presented a method based on PCA for detection and diagnosis of sensor failures where faults in AHU were isolated by Q-contribution plot and used squared prediction error as indices of fault detection. Likewise, PCA along with Joint Angle Analysis (JAA) is also proposed by Du et al. [10] for diagnosis of sensors' drifting and fixed biases in Variable Air Volume (VAV) systems. A new method for the detection of drifting biases of sensors in air handling unit is proposed by Du et al. [11] which employed neural networks along with wavelet analysis. Zhu et al. [32] adduced a sensor failure detection system based on regression neural network. It employed the analysis made by three-level wavelet for decomposition of the measured sensor data followed by extraction of each frequency band's fractal dimensions for the depiction of sensor's failure characteristics and then it is trained with neural networks to diagnose failures. A new semi supervised method for detection and diagnosis of air handling unit faults is proposed by Yan et al. [29] where a small amount of faulty training data samples were used to give the performance comparable to the classic supervised FDD methods. Madhikermi et al. [19] presented a heat recovery failure detection method in AHU using logistic regression and PCA. This method is based on process history and utilizes nominal efficiency of AHU for detection of faults.

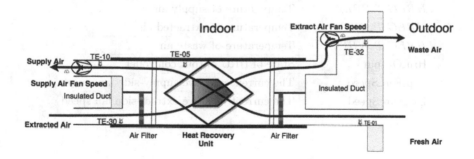

Fig. 3. The schematic diagram of Heat Recycler Unit

3 Theoretical Background

3.1 Heat Recycler Unit

A typical AHU with balanced air ventilation system, as shown in Fig. 3, includes the HRU, supply fan, extract fan, air filters, controllers, and sensors. The system circulates the fresh air from outside to the building by utilizing two fans (supply side and extract side) and two ducts (fresh air supply and exhaust vents). Fresh air supply and exhaust vents can be installed in every room, but typically this system is designed to supply fresh air to bedrooms and living rooms where occupants spend their most of time. A filter is employed to remove dust and pollen from outside air before pushing it into the house. The system also extracts air from rooms where moisture and pollutants are most often generated (e.g. kitchen and bathroom). One of the major component of the AHU is HRU which is used to save energy consumption. The principle behind the HRU is to extract heat from extracted air (before it is removed as waste air) from house and utilize it to heat fresh air that is entering into the house. HRU is a fundamental component of AHU which helps to recycle extracted heat. The main controllers included in the system are supply air temperature controller which adjusts the temperature of the supply air entering into house and HRU output which controls the heat recovery rate. In order to measure efficiency of HRU, five temperature sensors are installed in AHU which measure the temperature of circulating air at different part of AHU (detailed in Table 1). In addition to data from sensors, HRU control state, supply fan speed, and extract fan speed can be collected from system.

Table 1. Dataset description of Air handling unit sensors

Sensor name	Description
HREG_T_FRS	Temperature of fresh incoming air
HREG_T_SPLY_LTO	Temperature of supply air after HRU
HREG_T_SPLY	Temperature of supply air
HREG_T_EXT	Temperature of extracted air
HREG_T_WST	Temperature of waste air
Hru_Output	State of HRU output controller
Sup_Fan_Speed	The current effective supply-side fan speed
Ext_Fan_Speed	The current effective extract-side fan speed

3.2   Support Vector Machine

SVM is a supervised machine learning approach used for both type of problems classification as well as regression. But most of the time it is used to solve classification problems. In this technique we plot all features as a data point in dimensional space by using coordinate values. Then a hyperplane is created that can discriminate the two classes easily. The problem in linear SVM using linear algebra for assisting the learning of the hyperplane. The equation for predicting a new input in linear SVM is calculated by using dot product between the input (x) and each support vector (xi) given as [1]:

$$f(x) = B(0) + sum(ai \times (x, xi)) \tag{1}$$

This equation involves the calculation of the inner products of a new input vector (x) with all support vectors in training data. The learning algorithm's training data helps in estimation of the coefficients B0 and ai (for each input). SVM model used in the proposed methodology can be depicted in Fig. 4 where two classes (Since it is binary classification problem) are shown which depicts the normal cases and fault detection cases with no heat recovery.

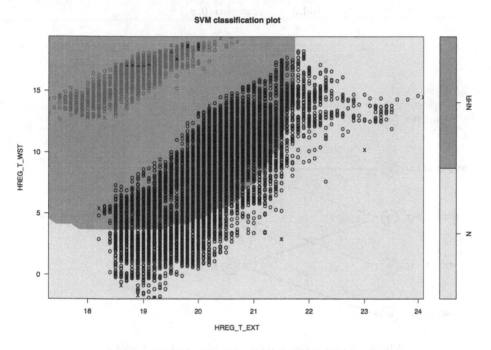

Fig. 4. SVM model used in proposed methodology

3.3   Neural Networks

Neural Networks the general function approximations, which makes them applicable to almost all machine learning problems where a complex mapping is to be learned from input to the output space. The computer based algorithms modeled on the behaviour and structure of human brain's neurons to train and categorize the complex patters are known as Artificial neural networks (ANNs). In artificial neural networks, the adjustment of parameters with the help of a process of minimization of error due to learning from experience leads to pattern recognition. The neural networks can be calibrated using different types of input data and the output can be categorized into any number of categories. The activation function can be used to restrict the value of output by squashing the output value and giving it in a particular range depending on the type of activation function used.

Table 2. The activation functions used in neural networks

Function	Formula
Sigmoid	$y_s = \frac{1}{1+e^{-x_s}}$
Tanh	$y_s = tanh(x_s)$
ReLu	$y_s = max(0, x_s)$

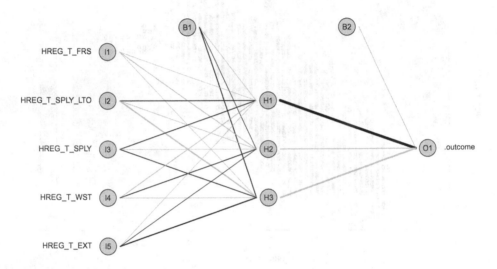

Fig. 5. Neural network model for proposed methodology

Table 2 lists the most common activation functions used in the neural networks where the value of sigmoid ranges from 0 to 1, tanh from -1 to 1 and ReLu from 0 to $+$infinity. Figure 5 depicts the neural network used in the proposed model which takes the five features described in the dataset as input which is mapped to hidden layers and finally to the output classifying it as fault detection or not.

3.4   Explainable Artificial Intelligence

Although there is an increasing number of works on interpretable and transparent machine learning algorithms, they are mostly intended for the technical users. Explanations for the end-user have been neglected in many usable and practical applications. Many researchers have applied the explainable framework to the decisions made by model for understanding the actions performed by a machine. There are many existing surveys for providing an entry point for learning key aspects for research relating to XAI [6]. Anjomshoae et al. [8] gives the systematic literature review for literature providing explanations about inter-agent explainability. The classification of the problems relating to explanation and black box have been addressed in a survey conducted by Guidotti et al. [15] which helped the researchers to find the more useful proposals. Machine learning models can be considered reliable but they lack in explainability. Contextual Importance and Utility has quite significance in explaining the machine learning models by giving the rules for machine learning models explanation [13]. Framling et al. provides the black box explanations for neural networks with the help of contextual importance utility [12, 14].

There are many methods used for providing the explanations for example; LIME (Local Interpretable Model-Agnostic Explanations) [3], CIU (Contextual Importance and Utility) [13], ELI5 [2], Skater [5], SHAP (SHapley Additive exPlanations) [4] etc. Most of them are the extensions of LIME which is an original framework and approach being proposed for model interpretation. These model interpretation techniques provide model prediction explanations with local interpretation, model prediction values with shape values, building interpretable models with surrogate tree based models and much more. Contextual Importance (CI) and Contextual Utility (CU) explains the prediction results without transforming the model into an interpretable one. These are numerical values represented as visuals and natural language form for presenting explanations for individual instances [13]. The CIU has been used by Anjomshoae et al. [7] to explain the classification and prediction results made by machine learning models for Iris dataset and car pricing dataset where the authors have CIU for justifying the decisions made by the models. The prediction results are explained by this method without being transformed into interpretable model. It explains the explanations for the linear as well as non linear models demonstrating the felexibility of the method.

4   Methodology

The proposed methodology considers the fact that due to high number of dimensions, detecting the failure cases (due to HRU failure) from the normal ones is really tedious task. The HRU's nominal efficiency (μ_{nom}) is a function of AHU's air temperatures as depicted in Eq. 2 [23]. The real dataset is collected from AHU containing 26700 instances of data collected for both states; "Normal" and "No Heat Recovery" state. There are two class labels with one label as "Normal" with 18882 instances and other as "No Heat Recovery" with 7818 instances. Since HRU output is set to "max" (i.e. it is a constant parameter) and HRU nominal efficiency being a function of air temperature associated with AHU (as shown in Eq. 2), this analysis only contains temperature differences as key point. All these dimensions have been combined together for measuring the performance of HRU.

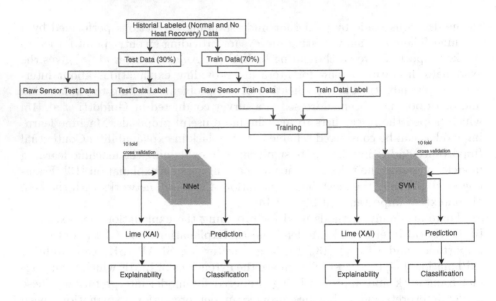

Fig. 6. Heat recovery failure detection methodology

$$\mu_{nom} = \frac{T_{ext} - T_{wst}}{T_{ext} - T_{frs}} \tag{2}$$

The methodology for the detection of heat recycler unit failure has been depicted in Fig. 6. The methodology starts with having the input data containing 5 features and 1 binary class label ("No Heat Recovery" or "Normal"). The input data is divided into 70:30 ratio of for training and testing dataset respectively. The training dataset is used for training 2 models neural networks (nnet) and support Vector Machine (SVM) individually along with 10 fold cross validation. After both the models have been trained on the training dataset, they both are tested for prediction on the testing dataset for the classification. Further, the justification for the decision made by both the models is given with the help of Explainable Artificial Intelligence (XAI). Local Interpretable Model-Agnostic Explanations (LIME) has been used for providing the explanation of both the models for 6 random instances of test data. The LIME helps in justifying the decisions made by the models, neural networks and SVM.

5   Result Analysis

The performance of the proposed methodology has been tested on two trained models, neural networks and support vector machine. The test dataset is given to both the trained models for obtaining the various performance metrics such as accuracy, sensitivity, specificity, precision, recall, confusion matrix and ROC. Table 3 compares the results obtained from both the models where neural networks outperforms the SVM. It shows that neural networks have the sensitivity and specificity as 0.91 and 1 respectively with accuracy of 0.97 whereas SVM has accuracy of 0.96 with sensitivity and specificity values as 0.99 and 0.95 respectively.

Table 3. Performance comparison of neural networks and SVM

Method	Accuracy	F1 score	Sensitivity	Specificity	Precision	Recall
Neural networks	0.97	1	0.90	1	1	0.90
SVM	0.96	1	0.99	.95	0.90	0.99

Table 4. Confusion matrix for nnet model

	No Heat Recovery	Normal
No Heat Recovery	2322	0
Normal	255	5463

Table 5. Confusion matrix for SVM model

	No Heat Recovery	Normal
No Heat Recovery	2364	255
Normal	21	5370

The confusion matrix obtained for neural networks and SVM is given in Tables 4 and 5 respectively. Here, the positive class is taken as 'No Heat Recovery' where there is failure in HRU and negative class is taken as 'Normal'. Table 4 shows that there are 2322 instances of True Positives (TP), 0 False Positives (FP), 255 False Negatives (FN) and 5463 True Negatives (TN) according to predictions made by neural network model. Similarly, Table 5 shows that there are 2364 instances of True Positives (TP), 255 False Positives (FP), 21 False Negatives (FN) and 5370 True Negatives (TN) according to predictions made by SVM model. ROC (Receiver Operating Characteristics) curve is one of the most important evaluation metrics for checking any classification model's performance. The ROC curve is used for diagnostic test evaluation where true positive rate (Sensitivity) is plotted as function of the false positive rate (100-Specificity) for different cut-off points of a parameter. The ROC curve for neural networks is depicted in Fig. 7 and for SVM is depicted in Fig. 8.

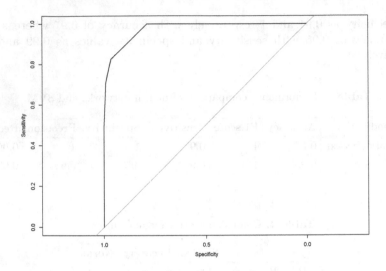

Fig. 7. ROC curve for Nnet model

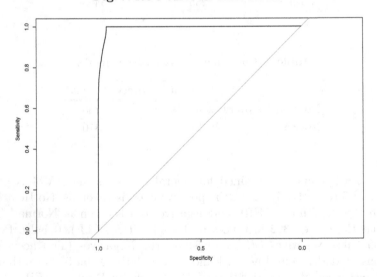

Fig. 8. ROC curve for SVM model

5.1   Explanations Using LIME

Since most of the machine learning models used for classifications or predictions are black boxes, but it is vital to understand the rationalization behind the predictions made by these machine learning models as it will of great benefit the decision makers to make the decision whether to trust the model or not. Figure 9 depicts an example of the case study considered in this paper for predicting the failure of the heat recovery unit. The explainer then explains the predictions made by the model by highlighting the causes or features that are critical in making the decisions made by model. However, it is possible that the model may make mistakes in predictions that are too hard to accept therefore, understanding model's predictions is quite important tool in deciding the trustworthiness of the model since the human intuition is hard to apprehend in evaluation metrics. Figure 10 illustrates the pick up step where convincing predictions are being selected for being explained to the human for decision making.

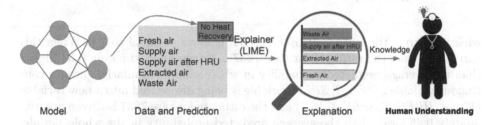

Fig. 9. Explaining individual predictions to a human decision-maker

Local Interpretable Model-agnostic Explanations (LIME) has been used for giving the explanations of the model which can be used by decision makers for justifying the model behaviour. The comprehensive objective of LIME is identifying an interpretable model over the interpretable representation which fits the classifier locally. The explanation is generated by the approximation of the underlying model by interpretable model which has learned on the disruptions of the original instance. The major intention underlying LIME is that it is being easier approximating black box model locally using simple model (locally in the neighbourhood of the instance) in contrast to approximating it on a global scale. It is achieved by weighing the original instances by their similarity to the case we wish to explain. Since the explanations should be model agnostic, LIME Because our goal should be to have model-agnostic model, We can use LIME for explaining a myriad of classifiers (such as Neural Networks, Support Vector Machines and Random Forests) in the domain of text as well as images [22].

The predictions made by both the models are then justified with the help of explainable artificial intelligence. Local Interpretable Model-Agnostic Explanations (LIME) has been used for providing the explanation of both the models for 6 random instances of test data. The explainability of neural networks and SVM is shown in Figs. 11 and 12 respectively. "Supports" means that the presence of that feature increases the probability for that particular instance to be of that particular class/label. "Contradicts" means that the presence of that feature decreases the probability for that particular instance to be of that particular class/label. "Explanation fit" refers to the R^2 of the model that is fitted locally to explain the variance in the neighbourhood of the examined case.

The numerical features are discretized internally by LIME. For instance, in Fig. 11, for case no. 7637, the continuous feature $HREG\_T\_WST$ is being discretized in such a way that a new variable is created ($HREG\_T\_WST \leq 7.1$) that

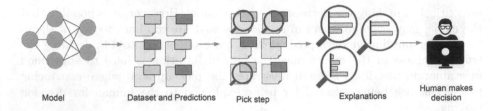

Fig. 10. Explaining the model to a human decision maker [22]

when it is true, the feature $HREG\_T\_WST$ is lower or equal to 7.1. When this variable is true, the estimate for case 7637 is driven approximately 0.34 higher than the average predicted probability in whole sample. Similarly, another continuous variable $HREG\_T\_SPLY$ variable is being discretized into a new variable ($12.9 < HREG\_T\_SPLY \leq 16.7$) and the estimate for case 7637 is driven approximately 0.45 lower than the average predicted probability in the whole sample, etc. When all the contributions are added on the average performance, it gives the final estimate. It also tells the class for which that particular instance belongs and how the probabilities of all variables have contributed in deciding that it belongs to that class. Similarly, it can be explained for second model SVM in Fig. 12.

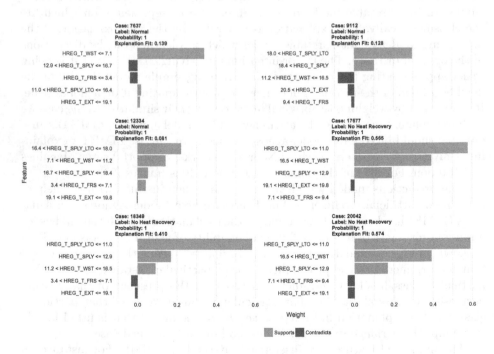

Fig. 11. Explainability of NNet model

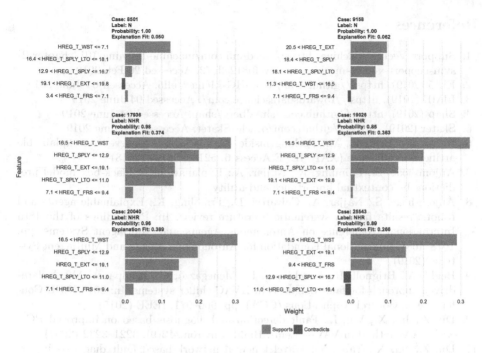

Fig. 12. Explainability of SVM

6   Conclusion

The heat recycler's fault detection in Air Handling Unit (AHU) is tedious task because the reason for its failure is mostly unknown and unique. The key requirement of such systems is the early diagnosis of such faults for its economic and functional efficiency. The real dataset of Heat Recycler Unit of AHU has been used for making predictions. The machine learning models, Support Vector Machine and Neural Networks have been used individually for the classification to detect the faults in AHU. Further, an explainable artificial intelligence has been used to explain the behavior of both the models i.e. the reason for justifying the recommendation or decision made by the learning models. Local Interpretable Model-Agnostic Explanations (LIME) has been used for providing the explanation of both the models chosen for 6 random instances of test data LIME has been used as an adequate tool for facilitating the trust for experts of machine learning and has been a good choice to be added in their tool belts. As a future work, we will like to compare the explanation results obtained by LIME with Contextual Importance (CI) and Contextual Utility (CU) to study how these two methods behave differently in context with providing the explanations.

References

1. Support Vector Machine. https://medium.com/machine-learning-101/chapter-2-svm-support-vector-machine-theory-f0812effc72. Accessed 26 Feb 2019
2. ELI5 (2019). https://github.com/TeamHG-Memex/eli5. Accessed 04 June 2019
3. LIME (2019). https://towardsdatascience.com/. Accessed 04 June 2019
4. Shap (2019). https://github.com/slundberg/shap. Accessed 04 June 2019
5. Skater (2019). https://github.com/oracle/Skater. Accessed 04 June 2019
6. Adadi, A., Berrada, M.: Peeking inside the black-box: a survey on explainable artificial intelligence (XAI). IEEE Access **6**, 52138–52160 (2018)
7. Anjomshoae, S., Främling, K., Najjar, A.: Explanations of black-box model predictions by contextual importance and utility
8. Anjomshoae, S., Najjar, A., Calvaresi, D., Främling, K.: Explainable agents and robots: results from a systematic literature review. In: Proceedings of the 18th International Conference on Autonomous Agents and Multiagent Systems, pp. 1078–1088. International Foundation for Autonomous Agents and Multiagent Systems (2019)
9. Beghi, A., Brignoli, R., Cecchinato, L., Menegazzo, G., Rampazzo, M.: A data-driven approach for fault diagnosis in HVAC chiller systems. In: 2015 IEEE Conference on Control Applications (CCA), pp. 966–971. IEEE (2015)
10. Du, Z., Jin, X., Wu, L.: Fault detection and diagnosis based on improved PCA with JAA method in VAV systems. Build. Environ. **42**(9), 3221–3232 (2007)
11. Du, Z., Jin, X., Yang, Y.: Wavelet neural network-based fault diagnosis in air-handling units. HVAC&R Res. **14**(6), 959–973 (2008)
12. Främling, K.: Explaining results of neural networks by contextual importance and utility. In: Proceedings of the AISB 1996 Conference. Citeseer (1996)
13. Främling, K.: Modélisation et apprentissage des préférences par réseaux de neurones pour l'aide à la décision multicritère. Ph.D. thesis, INSA de Lyon (1996)
14. Främling, K., Graillot, D.: Extracting explanations from neural networks. In: Proceedings of the ICANN, vol. 95, pp. 163–168. Citeseer (1995)
15. Guidotti, R., Monreale, A., Ruggieri, S., Turini, F., Giannotti, F., Pedreschi, D.: A survey of methods for explaining black box models. ACM Comput. Surv. (CSUR) **51**(5), 93 (2018)
16. Gunning, D.: Explainable artificial intelligence. Technical report released by DARPA (2017)
17. Holzinger, A., Biemann, C., Pattichis, C.S., Kell, D.B.: What do we need to build explainable AI systems for the medical domain? arXiv preprint arXiv:1712.09923 (2017)
18. Lee, W.Y., House, J.M., Park, C., Kelly, G.E.: Fault diagnosis of an air-handling unit using artificial neural networks. Trans.-Am. Soc. Heat. Refrig. Air Cond. Eng. **102**, 540–549 (1996)
19. Madhikermi, M., Yousefnezhad, N., Främling, K.: Heat recovery unit failure detection in air handling unit. In: Moon, I., Lee, G.M., Park, J., Kiritsis, D., von Cieminski, G. (eds.) APMS 2018. IAICT, vol. 536, pp. 343–350. Springer, Cham (2018). https://doi.org/10.1007/978-3-319-99707-0_43
20. Mills, E., et al.: The cost-effectiveness of commissioning new and existing commercial buildings: lessons from 224 buildings. In: Proceedings of the National Conference on Building Commissioning (2005)
21. Pérez-Lombard, L., Ortiz, J., Pout, C.: A review on buildings energy consumption information. Energy Build. **40**(3), 394–398 (2008)

22. Ribeiro, M.T., Singh, S., Guestrin, C.: Why should i trust you?: explaining the predictions of any classifier. In: Proceedings of the 22nd ACM SIGKDD International Conference on Knowledge Discovery and Data Mining, pp. 1135–1144. ACM (2016)
23. Roulet, C.A., Heidt, F., Foradini, F., Pibiri, M.C.: Real heat recovery with air handling units. Energy Build. **33**(5), 495–502 (2001)
24. Sheh, R., Monteath, I.: Introspectively assessing failures through explainable artificial intelligence. In: IROS Workshop on Introspective Methods for Reliable Autonomy (2017)
25. Van Lent, M., Fisher, W., Mancuso, M.: An explainable artificial intelligence system for small-unit tactical behavior. In: Proceedings of the National Conference on Artificial Intelligence, Menlo Park, CA, pp. 900–907. AAAI Press/MIT Press, Cambridge/London 1999 (2004)
26. Wang, S., Xiao, F.: Detection and diagnosis of ahu sensor faults using principal component analysis method. Energy Convers. Manag. **45**(17), 2667–2686 (2004)
27. Wang, X.F., Huang, D.S.: A novel density-based clustering framework by using level set method. IEEE Trans. Knowl. Data Eng. **21**(11), 1515–1531 (2009)
28. Xiao, F., Wang, S.: Progress and methodologies of lifecycle commissioning of HVAC systems to enhance building sustainability. Renew. Sustain. Energy Rev. **13**(5), 1144–1149 (2009)
29. Yan, K., Zhong, C., Ji, Z., Huang, J.: Semi-supervised learning for early detection and diagnosis of various air handling unit faults. Energy Build. **181**, 75–83 (2018)
30. Yan, R., Ma, Z., Kokogiannakis, G., Zhao, Y.: A sensor fault detection strategy for air handling units using cluster analysis. Autom. Constr. **70**, 77–88 (2016)
31. ten Zeldam, S., de Jong, A., Loendersloot, R., Tinga, T.: Automated failure diagnosis in aviation maintenance using explainable artificial intelligence (XAI). In: Proceedings of the European Conference of the PHM Society, vol. 4 (2018)
32. Zhu, Y., Jin, X., Du, Z.: Fault diagnosis for sensors in air handling unit based on neural network pre-processed by wavelet and fractal. Energy Build. **44**, 7–16 (2012)

Explainable Agent Simulations

Explaining Aggregate Behaviour in Cognitive Agent Simulations Using Explanation

Tobias Ahlbrecht[1]([✉])[iD] and Michael Winikoff[2][iD]

[1] TU Clausthal, Clausthal-Zellerfeld, Germany
tobias.ahlbrecht@tu-clausthal.de
[2] Victoria University of Wellington, Wellington, New Zealand
michael.winikoff@vuw.ac.nz

Abstract. We consider the problem of obtaining useful (and action-able) insight into the behaviour of agent-based simulation (using cognitive agents). When such simulations are being developed and refined, it can be useful to gain understanding of the simulation's behaviour. In particular, such understanding often needs to be *specific* to a given scenario (not just high-level generic information about the simulation dynamics), and about the *aggregate* behaviour of multiple agents. We describe a method for taking explanations of behaviour produced by individual agents, and aggregating them to obtain useful information about the aggregate behaviour of multiple agents. The method, which has been implemented, is illustrated in the context of a traffic simulation.

Keywords: Agent-based simulation · Explanation · Cognitive agents

1 Introduction

In this paper we show how explanation of agent behaviour can be used to gain understanding of the behaviour of a *collection* of agents. Specifically, we are interested in a collection of agents that do *not* operate as a team.

According to Malle [20], people use different kinds of explanations depending on whether a group is perceived as *jointly acting* or as an *aggregate* group. In the former case, the group coordinates its actions to achieve a joint goal. In the latter case - that we will approach in this paper - each agent acts individually and most often there is no explicit group, or it can only be ascribed after the fact. The result of these agents acting in a similar way is what we call *aggregate behaviour*.

One setting where the problem of understanding collective (aggregate) agent behaviour arises is in agent-based simulation. There are a number of different types of understanding that apply to such systems. One can be interested in

M. Winikoff—This paper was written while Michael was at the University of Otago.

D. Calvaresi et al. (Eds.): EXTRAAMAS 2019, LNAI 11763, pp. 129–146, 2019.
https://doi.org/10.1007/978-3-030-30391-4_8

understanding the high-level broad dynamics of a complex system. One can also be interested in understanding the detailed specific behaviour of an individual agent.

Our work sits *between* these two extremes: we are interested in understanding the behaviour of multiple agents, not just a single agent. But we are interested in understanding *specific* behaviour, rather than high-level dynamics.

This sort of understanding is typically needed as a simulation is developed and refined. For instance, in a traffic simulation, we might want to understand why, in a particular run, there was congestion on a particular road. This sort of understanding of a given simulation run can be of use in understanding the simulation, developing confidence that it is working correctly, and, if it is not, locating errors in the simulation behaviour.

We assume that the simulation uses *cognitive* agents, that is, agents that are conceptualised and implemented in terms of mental attitudes, such as beliefs, goals, and plans. These provide a basis for individual agents to explain their behaviour in terms of concepts that are familiar, since they are the same concepts that humans use to explain their behaviour [20].

Our key thesis is that we can then build on techniques for explaining the behaviour of cognitive agents in order to provide useful explanations of the *collective* behaviour of agents in a simulation.

Although there has been work on techniques for individual agents to explain their behaviour (e.g. [1,2,6,9,14,21,26]), there has not been work on understanding the collective behaviour of independent agents. However, we are not the first to propose using explanation for understanding collective agent behaviour in agent-based simulation. Harbers *et al.* [15] proposed this back in 2010. However, they did not provide a mechanism for doing this, or any details: their paper focused almost entirely on explaining the behaviour of a *single* cognitive agent.

In this paper we build on their work (and other more recent work on explaining the behaviour of cognitive agents [29]). Specifically, we propose a simple mechanism for aggregating multiple explanations, and a process for using this information to help obtain understanding of the behaviour of a simulation. We have implemented a simulation, including both individual agent explanation and aggregation of explanations. We use this simulation to show how the proposed mechanism can be helpful in gaining understanding of the behaviour of the simulation.

The remainder of this paper is structured as follows. We next (Sect. 2) briefly review background material on cognitive agents, and on techniques explaining the behaviour of cognitive agents. Section 3 presents the case study traffic simulation, including how the explanation mechanism is applied. Then (Sect. 4) we present the mechanism and process for aggregating and using individual agent explanations to help gain understanding. This is illustrated using the implemented simulation. Finally, we finish with a discussion including future directions (Sect. 5).

2   Background: BDI and Explanation

In this section, we briefly review some required background. We begin by introducing the notion of cognitive agent that we use, specifically the BDI model, and the AgentSpeak cognitive agent-oriented programming language. Then we review the approach for explaining cognitive agent behaviour that we use as a foundation.

In this paper we adopt the well-known Belief-Desire-Intention (BDI) model for cognitive agents [7,25]. This model, which has its roots in philosophical work drawing on the folk psychology of human decision making, views rational autonomous agents in terms of mental attitudes. Specifically, agents are viewed as having (or being ascribed) *plans*, *beliefs*, *desires*, and *intentions*[1]. Beliefs are information that the agent has about the environment or itself. Plans can be thought of as canned recipes for achieving a desired outcome, or responding to some sort of change. Desires are situations that the agent wants to bring about, and intentions are plan *instances* that the agent has decided upon as the means to achieve its desired ends.

The BDI model has been instantiated in a number of agent-oriented programming languages, including PRS [12,18] and AgentSpeak (and the Jason extension) [5,24,27], as well as GOAL [16], JACK [8,28], JAM [17], dMARS [11], UM-PRS [19], SPARK [22], Jadex [23], GWENDOLEN [10]), and others [3,4].

In this paper we use the AgentSpeak language, which was originally developed by Rao [24] as an abstraction of existing BDI languages at the time. An AgentSpeak agent has a set of beliefs (basically logical propositions) and a (static) set of plans. Each plan π_i is of the form $t_i : c_i \leftarrow b_i$ where t_i is a *trigger*, c_i a *context condition*, and b_i the plan body. A plan can be triggered by the addition or removal of a belief (denoted respectively $+b$ and $-b$), or by the addition or removal of a goal (respectively denoted $+!g$ and $-!g$). The context condition is a logical expression (usually restricted to conjunctions of atoms, where an atom is a proposition or a negated proposition). Finally, the plan body b_i is a sequence of steps $s_1; \ldots; s_n$. Possible steps are: adding a belief $+b$, removing a belief $-b$, updating a belief $-+b$, posting a sub-goal $!g$, or performing an action in the environment (`.action(params)`).

The semantics of AgentSpeak is that when a trigger occurs (e.g. a goal is added), the agent collects all the plans that can be used to handle that trigger (termed the *relevant* plans). It next computes the subset of the relevant plans for which the context condition c_i is currently true, termed the *applicable* plans. The agent then selects one of the applicable plans and executes it. This selection and execution process is interleaved with further event processing. In other words, it is done step by step, using a data structure (*intention stack*) to keep track of incremental execution of plans.

The original AgentSpeak paper leaves a number of details unspecified, such as the order in which plans are considered, and what failure recovery mechanism should be used if a step in a plan body fails. In this paper we follow the semantics

---

[1] Despite the acronym being "BDI", plans are an essential component.

of the AgentSpeak implementation that we use[2] by ignoring failure handling, and considering plans in the order in which they appear.

We now turn to how we can explain the behaviour of cognitive agents. Space precludes a detailed discussion, and we refer the reader to existing work cited below.

Our explanation mechanism is a simplified version of Winikoff et al. [29]. This can be used to generate explanation for the behaviour of *single* agents. The reminder of this section reviews their mechanism, whereas the next section describes a simulation domain, and how to apply the mechanism in this domain (which is novel).

This work is motivated by Miller's observation that explaining agent behaviour ought to draw on insights from the social sciences [21]. Specifically, we follow Malle [20], who argues that people use concepts of folk psychology, i.e. beliefs, desires and values to explain behaviour and furthermore [13], that autonomous systems should explain their behaviour in terms of the same concepts to deliver explanations that are *more comprehensible*. This aligns nicely with the use of BDI agents, which are conceptualised and implemented in terms of these same concepts.

In essence, the explanation mechanism that we will use (from [29]) takes (i) a *trace* of actions, (ii) the agent program (viewed as a goal-plan tree), and (iii) a query (a particular action). It then constructs a set of *explanatory factors* by traversing the tree. An explanatory factor can be a condition (occurring in an action's pre-condition or in the context condition of a plan), or a goal node[3]. The following gives a brief description of the process, for details we refer the reader to Winikoff et al. [29].

The process for constructing the explanation first prunes the trace by removing everything after the action being explained. The process then traverses the goal-plan tree from top to bottom, collecting explanatory factors using the following cases:

- Leaf (i.e. an action): if the action has a pre-condition then return its pre-condition, otherwise return the empty set
- Sequence (i.e. a plan body): collect the explanatory factors from all sub-trees which play a role in execution (formally: that contain an action node which appears in the pruned trace)
- Alternative (i.e. a goal with more than one relevant plan): collect the explanatory factors from all sub-trees which play a role in execution (same as the previous case). Then, if the action being explained appears in the sub-tree rooted at the current node, then call *pref* (see below) and add the results to the collected explanatory factors.

The function *pref* considers a choice point, and constructs an explanation for why the particular path taken was chosen. This explanation has two components,

---

[2] https://github.com/niklasf/pyson.

[3] There are also other types of explanatory factors in the paper that we do not use here, namely "tried-but-failed", "valuings", and "forward links".

where we assume that plan π_i with context condition c_i was selected for execution, and plans π_j $(1 \leq j < i)$ were therefore not selected:

1. the context condition of the selected plan must be true for it to be taken, so the explanation includes this condition c_i; and
2. each of the other options (i.e. plans) that appear earlier in the program (and hence are considered false), have a false context condition, so we include this, i.e. $\neg c_j$ for all c_j.

3 Running Example: Traffic Simulation

For our first simulation trial, we chose the traffic domain as it is very descriptive and allows for many basic actors in a shared environment. Results of a traffic simulation might be usage statistics for each road, so they could be used to predict bottlenecks in what-if scenarios. For example, one might be interested in why a specific road was particularly congested during a certain time frame. As we are interested in the general feasibility of our approach, we do not require truly realistic behaviour or outcomes. Thus, we implemented a simple simulation model that contains only a few key elements and simplistic agent behaviour. Nevertheless, it may generate overall simulation behaviour that is complex to explain without the right support mechanism.

The environment consists of a simple (undirected) road network where roads are represented as edges and intersections as the nodes between them. Each road has a certain length and a dynamic traffic level that depends on the number of agents currently using the road. In addition, each road can also be a bridge, which is either open or closed, effectively rendering the road unusable for a certain amount of time. This allows for agents with different preferences choosing different routes.

Each agent's goal is to go from one node in the network to another. We abstract away from actual fine-grained driving mechanics and focus solely on the navigation-related decision-making of agents. First, if agents encounter a bridge, they may prefer to wait until it opens, or take a detour. Also, they check the current traffic on the next roads they can take, and if it is too heavy, they might decide to take a longer route if they expect it to have less traffic.

The agent program has been implemented in AgentSpeak[4]. The simulation triggers the agent program by adding a new *step* belief during each simulation cycle and waits until all agents have nothing left to reason about. While the simulation program by default generates a random graph, and random agent preferences, it can also be given a specific setup, such as that shown in Fig. 1.

We now briefly describe the percepts and beliefs, the actions, and the decision-making logic of the agents.

The agents receive[5] the following beliefs about the structure of the environment and the behaviour they prefer:

---

[4] Full source code is available from https://github.com/t-ah/group-explanations.

[5] i.e. these beliefs are set and updated by the simulation program.

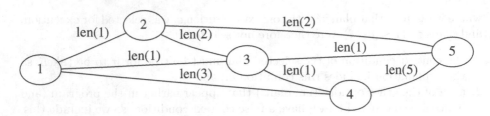

Fig. 1. Simple road network where annotations on an edge indicate its length (len).

- `step`: tells the agent a new simulation step has begun,
- `node(N)`: there is a node/intersection called N,
- `edge(N1, N2, L, _)`: there is a road from N1 to N2 of length L,
- `bridge(N1, N2)`: the road from N1 to N2 is a bridge,
- `position(P)`: the agent's location; either the name of a node or a list containing 2 nodes describing the road between them,
- `destination(D)`: the agent wants to reach node D, and
- `waitForBridges`: the agent prefers to wait if a bridge is closed.

The agents also use a number of beliefs to keep track of what they have already done and what they want to do in the medium term:

- `plannedRoute(R)`: the agent wants to follow the route R (a list of nodes), and
- `usedRoad(A,B)`: the agent has already tried the road from A to B.

The following shows the actions that are defined in the simulator, and their pre- and post-conditions.

.nextSteps input: position & destination, output: a sorted list of terms each of the form $road(neighbour, path\_length)$ giving a possible neighbour to the current node, and the path length to the destination. No pre- or post-conditions (other than a path existing).

.takeRoad inputs: node & nextNode, no outputs. Pre-condition: there is a road from node to nextNode; and if the road is a bridge, it is open or the agent is already on this road. Post-condition: progresses the agent on the road or updates the position to nextNode.

.bridgeStatus input: node1 & node2, output: is the bridge open? Pre-condition: the edge from node1 to node2 is a bridge, and its status is recorded.

.getDetour input: target, output: a path from the current position to the specified target, assuming the edge from position to target is given a high cost (1000). Pre-condition: there is an edge from the current position to the target.

.getTraffic input: target, output: the additional amount of steps it will take the agent to traverse the road to the target. Traffic is dependent on the number of agents on a road. The more agents there are on a road, the less progress an agent makes on that road per step.

$$Progress = 0.2 + ((1 - 0.2)/(\#_{agents}(road) + 1))$$

.logStep input: item to be logged. No pre- or post-conditions, or output. This action logs information to the trace to be presented to the user.

Finally, each agent uses the decision-making logic presented in Algorithm 1.

Algorithm 1. Agent Decision-Making Logic

1: $ag \leftarrow$ the agent; $pos \leftarrow position(ag)$; $dest \leftarrow destination(ag)$
2: **if** ag on $road(X, Y)$ **then**
3:     $takeRoad(X, Y)$ ▷ continue using road
4: **else**
5:     **if** $pos = dest$ **then**
6:         stop
7:     **else if** $PlannedRoute \neq \emptyset$ **then**
8:         $node_{selected} \leftarrow$ next element of $PlannedRoute$
9:     **else**
10:         $N_{nextSteps} \leftarrow neighbours(pos)$
11:         $N_{unused} \leftarrow \{n \mid n \in N_{nextSteps} \wedge \neg usedRoad(pos, n)\}$
12:         **if** $N_{unused} = \emptyset$ **then**
13:             $N_{unused} = N_{nextSteps}$
14:         $N_{best} = \{n \mid n \in N_{unused} \wedge \forall n' \in N_{unused} :$
                  $len(pos, n) + getTraffic(pos, n) \leq len(pos, n') + getTraffic(pos, n')\}$
15:         $node_{selected} \leftarrow N \in N_{best}$
16:         **if** $node_{selected}$ is a bridge and currently closed **then**
17:             **if** ag prefers to wait at bridges **then** wait
18:             **else**
19:                 $PlannedRoute \leftarrow planDetour(pos, node_{selected})$
20:                 $node_{selected} \leftarrow$ next element of $PlannedRoute$
21:     $takeRoad(pos, node_{selected})$ ▷ switch to the selected road

---

3.1 Explaining Decisions

We now describe how we apply the explanation mechanism from Winikoff *et al.* [29] (briefly summarised in Sect. 2) in this scenario.

Recall that an explanation is a set of explanatory factors, with each explanatory factor being a desire or a condition. An explanatory factor condition can be an action's pre-condition, or a complex formula that explains why a particular plan was selected (basically a combination of that plan's context condition being true, and the context conditions of other relevant plans being false; and in this case, where plans are considered in the order in which they appear in the program, "other plans" means "plans appearing earlier in the program").

For this work we manually modify the agent program so that it captures elements of the explanation on-the-fly. This is done by adding at key points (see below) logging statements that capture the relevant explanatory factors. The logging statements build up a trace data structure that then can be used after execution to provide explanations for questions of the form "why did agent A_i perform a particular action?".

Capturing explanatory factors that are desires is done by firstly identifying which desires (i.e. goals) we wish to record, and then, for each plan that is triggered by the posting of one of these goals, we simply add to the start of the plan a logging statement that records the trigger of that plan.

We now turn to capturing explanatory factors that are conditions. There are two cases of condition explanatory factors being captured.

The first, and simple one, is that just before an action is performed, we check that the action's pre-condition holds[6], and log that action's pre-condition as an explanatory factor before performing the action. We also add a step to log that the action was performed.

The second, more complex, case concerns selecting which plan to use next. Whenever a trigger has more than one plan, then we add a first step to each plan that logs the explanatory factors. For the first plan (for that trigger), the explanatory factor is the plan's context condition. For subsequent plans (for that trigger) the explanatory factor is that plan's context condition, as well as the negation of earlier plans' context conditions.

For example, given the following plans:

```
+!goto(To) : position(Pos) ∧ bridge(Pos, To) ∧
.bridgeStatus(Pos, To, open(false)) ∧ waitForBridges. // wait
+!goto(To) : position(Pos) ∧ bridge(Pos, To) ∧
.bridgeStatus(Pos, To, open(false)) ∧ not plannedRoute(_) ←
    !useDetour(To).
+!goto(To) : position(Pos) ∧ bridge(Pos, To) ←
    +usedRoad(Pos, To); .takeRoad(Pos, To).
+!goto(To) : position(Pos) ←
    +usedRoad(Pos, To); .takeRoad(Pos, To).
```

We modify them by adding logging steps as follows (combining some factors for readability):

```
+!goto(To) : position(Pos) ∧ bridge(Pos, To) ∧
.bridgeStatus(Pos, To, open(false)) ∧ waitForBridges ←
    .logStep(explain(goto(To), waitForClosedBridge(Pos, To))).
+!goto(To) : position(Pos) ∧ bridge(Pos, To) ∧
.bridgeStatus(Pos, To, open(false)) ∧ not plannedRoute(_) ←
    .logStep(explain(goto(To), notWaitForClosedBridge(Pos, To)));
    !useDetour(To).
+!goto(To) : position(Pos) ∧ bridge(Pos, To) ←
    .logStep(explain(goto(To), bridgeOpen(Pos, To)));
    +usedRoad(Pos, To); .logStep(action(takeRoad(Pos,To))); .takeRoad(Pos, To).
+!goto(To) : position(Pos) ←
```

---

[6] In this particular scenario this is not required, since action pre-conditions are either trivial, guaranteed by construction (e.g. a detour will only be attempted around an existing road), or are checked elsewhere (e.g. in context conditions).

```
.logStep(explain(goto(To), noBridge(Pos, To)));
+usedRoad(Pos, To); .logStep(action(takeRoad(Pos,To))); .takeRoad(Pos, To).
```

We then construct an explanation for a query as follows. A query is an action instance ("why was this action done?"). We construct the answer by finding the action instance in the trace, removing everything that occurs after it, and then collecting the explanatory factors in the remaining trace.

For example, considering the example scenario in Fig. 1, a car might start out at node 1 and desire to reach node 5. This car might start by taking the road to node 2, and then proceed to 5. We might wonder why the car chose to go via node 2, since the path via node 3 is shorter. Asking the query "why did this car take the road from node 1 to node 2?" results in the collection of explanatory factors below[7]. The relevant one is the third, which shows that it preferred the road from 1 to 2 over the road from 1 to 3 because there was traffic on the 1–3 road. The other factors below are that the agent had the goal of reaching node 5, that it initially had no planned route and was not at its destination (so it planned a route), that it preferred 1–2 over 1–4 because the path was shorter, and that it was able to take the road 1–2 because the road was not a bridge (if it was a bridge, then the condition would have been that the bridge was open).

```
Explanation:
reach(5)
notAtDestination & noPlannedRoute
would_prefer_due_to_traffic([1, 2], [1, 3])
would_prefer_due_to_route_length([1, 2], [1, 4])
goto(2), noBridge(1, 2)
```

Please note that we chose this approach of explicitly logging explanatory factors because for now it is a simpler solution for testing purposes. One of the next steps will be to move this functionality into the AgentSpeak interpreter itself, so that arbitrary agent programs can be handled without the agent programmer having to add any additional code related to the explanation mechanism.

4   Aggregation and Simulation Explanation Process

Now that we have applied the single agent explanation mechanism to our scenario, we are able to generate explanations for each agent's individual decisions. Next, we want to use these explanations to explain the behaviour of many agents at once. Thus, we need ways to ask the system questions, find relevant agents for these questions, and aggregate the explanations collected for each individual agent. This allows us to get explanations for the collective actions of groups of agents, including emergent phenomena.

A straightforward way to aggregate explanations is to count the occurrences of all explanatory factors that are related to a query, and list the most common ones. The general *aggregation process* looks as follows:

---

[7] The text "`explain(...)`" has been elided for readability.

1. Given a simulation run with agents $\mathcal{A} = \{A_1, \ldots, A_n\}$, and a query Φ, find the set of agents $\mathcal{A}^\Phi$ for which the query makes sense (e.g. if the query is "why did you take road R12?", then select the agents that did actually take that road).
2. For each $A_i \in \mathcal{A}^\Phi$ generate an explanation set E_i^ϕ which explains why agent i did ϕ (using the mechanism from [29]).
3. Identify factors that appear in explanations, along with their count:

$$\{(c, n^c) \mid \exists E_i^\phi : c \in E_i^\phi \wedge n^c = |\{A_i \mid c \in E_j^\phi\}|\}$$

In other words, for each factor c that appears in some explanation E_i, include that factor along with n^c. We define n^c as the number[8] of agents where c appears in that agent's explanation for the query ϕ.

The first assumption we want to test is, whether an explanatory factor that occurs more often is more suitable to be included in an explanation of the group's behaviour.

Steps 1–3 above are a computational process which can (and has been) implemented. The way in which these explanations are used is part of a larger human-driven process:

A. Identify an aspect of the simulation's behaviour that is interesting or surprising and pose a question.
B. Invoke the above steps (1–3) to obtain an answer.
C. In some situations an answer might include further potential questions. For example, it may be that an agent chose a particular road R_1 because there was traffic on another road R_2, which might then lead to the followup question "why did the agents take road R_2?". In this case, the human would pose the followup question and return to step B.
D. In other situations, an answer might be testable by modifying the simulation. For example, if many or most agents performing action A did so because of a condition c or some parameters, then we might modify c (or the parameters) and re-run the simulation to check whether in fact condition c is in itself a sufficient explanation. This ability to test explanations by performing counter-factual experiments relies on the setting of the work being simulation.

We now describe an example simulation that illustrates the process, and the way in which it can help to obtain understanding of a simulation.

4.1   Example Simulation

In this example, we have 5 nodes connected by 8 roads and 100 agents located at node 1 who want to reach node 5 (see Fig. 2). Half (50%) of the agents prefer to wait at closed bridges while the other 50% prefer to take a detour, rather than wait.

---

[8] The notation $|S|$ denotes the size (number of elements) of the set S.

Looking at the simulation results (left side of Fig. 2), we observe that quite a few cars take the road from 2 to 5, and ask why that road was taken? Note that from Fig. 2 we can tell that most of the cars that took the road from 2 to 5 came from 3. However, we cannot tell why they chose to go to 2, rather than proceed from 3 directly to 5. It could be due to traffic on the 3–5 road, or due to the bridge being closed.

The query gives the following list of explanatory factors (produced by the implementation). For now, we will look at and analyse the "raw" output of the explanation and aggregation method to see where it helps and where we can improve it in the future.

Note that "reach(5)" below indicates that the agent had the goal of reaching node 5. The explanatory factor "would prefer due to route length" indicates that the first option has a shorter path than the second, and that this remains the case when the agent takes into account known traffic.

```
Factors for (2, 5):
57 times noPlannedRoute
57 times noBridge(2, 5)
57 times notAtDestination
57 times reach(5)
53 times would_prefer_due_to_route_length([1, 3], [1, 4])
53 times noBridge(3, 2)
53 times would_prefer_due_to_route_length([3, 5], [3, 4])
53 times notWaitForClosedBridge(3, 5)
53 times would_prefer_due_to_route_length([3, 5], [3, 2])
53 times notAtDestination(2)
53 times useDetourAround(3, 5)
53 times would_prefer_due_to_route_length([3, 5], [3, 1])
53 times plannedRoute([5|[]])
53 times tookRoad(1, 3)
53 times noBridge(1, 3)
53 times tookRoad(3, 2)
50 times would_prefer_due_to_route_length([1, 3], [1, 2])
 7 times would_prefer_due_to_route_length([1, 2], [1, 4])
 7 times tookRoad(1, 2)
 7 times noBridge(1, 2)
```

Each line represents one of the 20 most common explanatory factors that were given by the agents that took the road from 2 to 5. Each factor is only counted once per agent, so for example 53 of the 57 agents who took the road came from node 3 and prefer to not wait for closed bridges to open, suggesting that they were trying to circumvent the bridge on road 3 → 5. This also indicates that traffic was not a factor, as none of the explanations are based on traffic and suggests that if we adjust the simulation to keep the bridge open more often, then this link would become less used. This hypothesis was therefore tested by re-running the simulation with adjusted bridge probabilities[9]. The results are on the right side of Fig. 2 and show the expected difference: traffic has moved from $2 \rightarrow 5$ to $3 \rightarrow 5$.

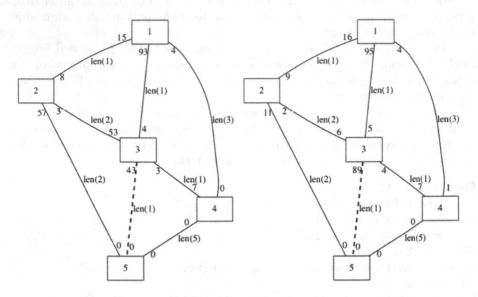

Fig. 2. Simulation Results: a number on an edge close to a node indicates the number of times a car took that edge from the node. For example, "4" near node 1 shows that 4 cars took the road from 1 to 4. Dashed edges are bridges. Left is original simulation, right is the re-run after adjusting the bridge probability.

Generally, we note that there are both more insightful factors and some less interesting ones. Looking at the frequencies of factors, we see that they can be a good first filter. Only the factors occurring in the top segment actually contribute to a useful explanation. On the other hand, taking all of these factors (which get a relatively high count) would create an explanation that is far too big. Thus, additional filtering will be necessary. For example, factors including the road that is currently being investigated are normally more relevant. Roads that are further away have overall less impact. We could change our explanation method to only collect explanatory factors from the current simulation step, however, that would already hide where the 53 agents came from. In other scenarios, it could be even more difficult to find a suitable cut-off point for the "age" of explanatory factors.

Another example of where aggregated explanations can offer insight is on the link between nodes 3 and 4. Let us begin with the $3 \rightarrow 4$ direction. At first glance, it is not clear why a small number of cars take this link. It might be

---

[9] The probability of an open bridge closing was changed from 0.6 to 0.4, and the probability of a closed bridge opening changed from 0.3 to 0.6.

because they encounter a closed bridge, and do not wish to wait. The aggregated explanation (below) shows that for all 3 cars, they actually are happy to wait for the closed bridge, and originally prefer $3 \rightarrow 5$, but increased traffic on the bridge (last line, possibly from immediately after it opens) drove them to take a detour.

We now turn to the $4 \rightarrow 3$ direction, to try and understand why cars might take this road. We might expect (from examining Fig. 2) that the cars in question chose to proceed from 1 to 4 due to traffic, and that, having arrived at 4, they prefer to go via 3, since it is shorter. However, we see from the aggregated explanation below that traffic did not play a role in the decision. This raises the question of why the agents took the road from 1 to 4. We also observe that of the 7 cars going from 4 to 3, only 4 came from 1. We therefore examine the precise paths taken by these 7 cars.

Looking at the seven cars we observe that the routes taken were: 1, 3, 4, 3, 5 (car 57); 1, 3, 4, 3, 2, 5 (cars 18 and 26); 1, 4, 3, 4, 1, 3, 5 (car 29); 1, 3, 1, 2, 1, 4, 3, 2, 5 (cars 25 and 55); and 1, 2, 1, 3, 1, 4, 3, 2, 5 (car 89).

In other words, the explanation (ruling out traffic as the reason for arriving at 4) has led us to examine the details of seven particular cars, which has allowed us to find odd (and unrealistic!) behaviour that is produced by the simulation. The reason for this odd behaviour is that the decision process considers traffic on the next link only. So, for example, when a car reaches node 3, with the plan of continuing directly to node 5, it then checks the traffic on the road from 3 to 5. If the traffic is high, then it may end up comparing the length from 3 to 5 directly (with high traffic, so slow), against the path from 3, to 1, then to 3, and then to 5. But this comparison only considers the traffic on the first link (from 3 back to 1), so, if there is very high traffic on 3-to-5, it might appear as a viable alternative to going directly from 3 to 5.

```
Factors for (3, 4):
3 times would_prefer_due_to_route_length([1, 3], [1, 2])
3 times would_prefer_due_to_route_length([3, 4], [3, 2])
3 times would_prefer_due_to_route_length([3, 5], [3, 1])
3 times would_prefer_due_to_route_length([3, 5], [3, 2])
3 times noPlannedRoute
3 times noBridge
3 times notAtDestination
3 times waitForBridges
3 times reach(5)
3 times would_prefer_due_to_traffic([3, 4], [3, 5])

Factors for (4, 3):
7 times noPlannedRoute
7 times noBridge
7 times would_prefer_due_to_route_length([4, 3], [4, 1])
7 times notAtDestination
7 times reach(5)
```

```
7 times would_prefer_due_to_route_length([4, 3], [4, 5])
6 times would_prefer_due_to_route_length([1, 3], [1, 2])
6 times would_prefer_due_to_route_length([3, 5], [3, 1])
6 times would_prefer_due_to_route_length([3, 5], [3, 2])
6 times waitForBridges
6 times would_prefer_due_to_route_length([3, 5], [3, 4])
6 times closedBridge
6 times would_prefer_due_to_route_length([1, 3], [1, 4])
```

We updated the simulation to address this issue, by improving how traffic is considered by the decision-making logic. First, each agent was modified to memorise all the traffic levels it has seen before, updating them with the actual traffic the agent can see on all incident roads whenever a node is reached. Second, the **.getTraffic** action was modified to return the length of the route to the agent's destination through a given target node, considering the agent's complete traffic knowledge. Having changed the behaviour to be more realistic with respect to dealing with traffic, we re-ran the simulation, giving the results in Fig. 3. We see now that agents do not backtrack anymore, which is more realistic. Also the longest route now includes only 3 roads (1, 3, 2, 5).

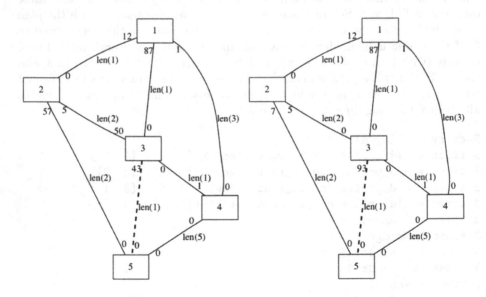

Fig. 3. Simulation Results after the fix

It is worth emphasising that the specific lessons about this simulation are not the interesting thing. The simulation here is an example of how aggregated explanations can be used to obtain useful (and actionable) insight about aggregate behaviour, at a level that is more detailed than just looking at the numbers.

For instance, looking at the original simulation we could see that many cars followed the path from 3 to 2 to 5, but we did not know whether this was due to traffic congestion, or due to some cars' policy of not waiting for a closed bridge. Additionally, we experienced first-hand how aggregated explanations may point to undesirable agent behaviour and help to find its origin.

4.2   Question Types

For now, the process is limited to explaining aggregated behaviour that is composed of agents performing similar actions.

In addition to having queries that request an explanation for an action being performed (e.g. "Why did the agents perform action X?"), we also want to permit queries that provide a condition (e.g. "Why did the agents perform action X under condition C?"). This can be handled by filtering (in the first step) to take into account the condition as well: we just have to filter the set of actions by checking the condition first.

Finally, we also want to allow queries that are based on a situation rather than an action, e.g. "Why are agents in situation S?". This is slightly more complex (and we do not cover it in detail in this paper). It can be handled by finding agents that were in the queried state at some point, and for each such agent, identifying the actions that led to it being in the queried states, and then requesting an explanation for those actions. Note that different agents might reach a state S via different actions, and so each agent might be queried with a different action to be explained. For our example, a question might be "why are the agents located at node N during steps 4–10?". Of course, in our scenario it is obvious that the action to take the road to node N led agents to be at N, so it is actually this action that is being queried. In other scenarios, it could also make sense to allow questions related to certain outcomes, if these outcomes can be associated with specific actions.

The explanation mechanism could also be helpful when trying to explain emergent behaviour, at least if such a behaviour can be retraced to the responsible actions. For example, in our simple simulation, we can observe (and explain) the emergent situation of road congestion.

5   Discussion

We have introduced a method for aggregating action explanations of individual agents to obtain useful information about their behaviour and outlined a process for using the method to gain insight into the behaviour of an agent-based simulation. Then we demonstrated the approach in the context of a simplified traffic simulation.

While we have already seen promising results, there are many things left to do.

First, we want to conduct a human participant evaluation to analyse the method's usability and effectiveness, especially regarding lay people who may not have a background in simulation and/or agent technology. For this, we will also need to improve the presentation of aggregated explanations. In particular, it would be interesting to investigate different presentations of explanations, e.g. as text, as dialogue, or in a graphical format.

Then, there are a few ways to improve the aggregation method that we need to investigate. For example, it might pay off to further aggregate or relate explanatory factors that are similar to each other, either because they were mentioned in the same explanation, or by the same agent. Also, we have seen that the relevance of an (aggregated) explanatory factor still depends on the specific query and we could devise a ranking method beyond (but not without) a factor's percentage. We could also investigate extending the explanation generation to incorporate causal reasoning, given domain knowledge.

Finally, this work only targets aggregated agent behaviour. On the other hand, there is also behaviour of jointly acting groups, that has to be discovered and explained in a different way. Integrating explanations for both kinds of group behaviour will lead to a more comprehensive way of explaining multi-agent behaviour and agent-based simulations in particular.

References

1. Aha, D., Darrell, T., Doherty, P., Magazzeni, D. (eds.): Proceedings of IJCAI 2018 Workshop on Explainable AI (XAI) (2018)
2. Aha, D., Darrell, T., Pazzani, M., Reid, D., Sammut, C., Stone, P. (eds.): Proceedings of IJCAI 2017 Workshop on Explainable AI (XAI) (2017)
3. Bordini, R.H., Dastani, M., Dix, J., El Fallah Seghrouchni, A. (eds.): Multi-Agent Programming: Languages, Platforms and Applications, vol. 15. Springer, Boston (2005). https://doi.org/10.1007/b137449
4. El Fallah Seghrouchni, A., Dix, J., Dastani, M., Bordini, R.H. (eds.): Multi-Agent Programming: Languages, Tools, and Applications. Springer, Boston (2009). https://doi.org/10.1007/978-0-387-89299-3
5. Bordini, R.H., Hübner, J.F., Wooldridge, M.: Programming Multi-Agent Systems in AgentSpeak Using Jason. Wiley, Hoboken (2007). ISBN 0470029005
6. Borgo, R., Cashmore, M., Magazzeni, D.: Towards providing explanations for AI planner decisions. In: Aha, D., Darrell, T., Doherty, P., Magazzeni, D. (eds.) Proceedings of IJCAI 2018 Workshop on Explainable AI (XAI), pp. 11–18 (2018)
7. Bratman, M.E.: Intentions, Plans, and Practical Reason. Harvard University Press, Cambridge (1987)
8. Busetta, P., Rönnquist, R., Hodgson, A., Lucas, A.: JACK intelligent agents - components for intelligent agents in Java. Technical report. Agent Oriented Software Pty. Ltd., Melbourne, Australia (1998). http://www.agent-software.com
9. Chakraborti, T., Sreedharan, S., Zhang, Y., Kambhampati, S.: Plan explanations as model reconciliation: Moving beyond explanation as soliloquy. In: Proceedings of the Twenty-Sixth International Joint Conference on Artificial Intelligence, IJCAI 2017, pp. 156–163 (2017). https://doi.org/10.24963/ijcai.2017/23
10. Dennis, L.A.: Gwendolen semantics: 2017. Technical report ULCS-17-001. University of Liverpool, Department of Computer Science (2017)

11. d'Inverno, M., Kinny, D., Luck, M., Wooldridge, M.: A formal specification of dMARS. In: Singh, M.P., Rao, A., Wooldridge, M.J. (eds.) ATAL 1997. LNCS, vol. 1365, pp. 155–176. Springer, Heidelberg (1998). https://doi.org/10.1007/BFb0026757

12. Georgeff, M.P., Lansky, A.L.: Procedural knowledge. Proc. IEEE **74**, 1383–1398 (1986). Special Issue on Knowledge Representation

13. de Graaf, M.M., Malle, B.F.: How people explain action (and autonomous intelligent systems should too). In: 2017 AAAI Fall Symposium Series (2017)

14. Harbers, M.: Explaining agent behavior in virtual training. SIKS dissertation series no. 2011-35. SIKS (Dutch Research School for Information and Knowledge Systems) (2011)

15. Harbers, M., Meyer, J.J., van den Bosch, K.: Explaining simulations through self explaining agents. J. Artif. Soc. Soc. Simul. (JASSS) **13**(1), 4 (2010). https://doi.org/10.18564/jasss.1437

16. Hindriks, K.V., de Boer, F.S., van der Hoek, W., Meyer, J.-J.C.: Agent programming with declarative goals. In: Castelfranchi, C., Lespérance, Y. (eds.) ATAL 2000. LNCS (LNAI), vol. 1986, pp. 228–243. Springer, Heidelberg (2001). https://doi.org/10.1007/3-540-44631-1_16

17. Huber, M.J.: JAM: a BDI-theoretic mobile agent architecture. In: Proceedings of the Third International Conference on Autonomous Agents (Agents 1999), pp. 236–243. ACM Press (1999)

18. Ingrand, F.F., Georgeff, M.P., Rao, A.S.: An architecture for real-time reasoning and system control. IEEE Expert **7**(6), 34–44 (1992)

19. Lee, J., Huber, M.J., Kenny, P.G., Durfee, E.H.: UM-PRS: an implementation of the procedural reasoning system for multirobot applications. In: Proceedings of the Conference on Intelligent Robotics in Field, Factory, Service, and Space (CIRFFSS 1994), pp. 842–849 (1994)

20. Malle, B.F.: How the Mind Explains Behavior. MIT Press, Cambridge (2004). ISBN 9780262134453

21. Miller, T.: Explanation in artificial intelligence: insights from the social sciences. Artif. Intell. **267**, 1–38 (2019). https://doi.org/10.1145/1824760.1824761

22. Morley, D., Myers, K.: The SPARK agent framework. In: Proceedings of the Third International Joint Conference on Autonomous Agents and Multiagent Systems (AAMAS), pp. 714–721. IEEE Computer Society, Washington, DC (2004)

23. Pokahr, A., Braubach, L., Lamersdorf, W.: Jadex: a BDI reasoning engine. In: Bordini, R.H., Dastani, M., Dix, J., El Fallah Seghrouchni, A. (eds.) Multi-Agent Programming. MSASSO, vol. 15, pp. 149–174. Springer, Boston, MA (2005). https://doi.org/10.1007/0-387-26350-0_6

24. Rao, A.S.: AgentSpeak(L): BDI agents speak out in a logical computable language. In: Van de Velde, W., Perram, J.W. (eds.) MAAMAW 1996. LNCS, vol. 1038, pp. 42–55. Springer, Heidelberg (1996). https://doi.org/10.1007/BFb0031845

25. Rao, A.S., Georgeff, M.P.: An abstract architecture for rational agents. In: Rich, C., Swartout, W., Nebel, B. (eds.) Proceedings of the Third International Conference on Principles of Knowledge Representation and Reasoning, pp. 439–449. Morgan Kaufmann Publishers, San Mateo (1992)

26. Roth-Berghofer, T., Schulz, S., Bahls, D., Leake, D.B. (eds.): Explanation-Aware Computing, Papers from the 2007 AAAI Workshop, Vancouver, British Columbia, Canada, 22–23 July 2007, AAAI Technical report, vol. WS-07-06. AAAI Press (2007)

27. Winikoff, M.: An AgentSpeak meta-interpreter and its applications. In: Bordini, R.H., Dastani, M.M., Dix, J., El Fallah Seghrouchni, A. (eds.) ProMAS 2005. LNCS, vol. 3862, pp. 123–138. Springer, Heidelberg (2006). https://doi.org/10.1007/11678823_8

28. Winikoff, M.: Jack™ intelligent agents: an industrial strength platform. In: Bordini, R.H., Dastani, M., Dix, J., El Fallah Seghrouchni, A. (eds.) Multi-Agent Programming. MSASSO, vol. 15, pp. 175–193. Springer, Boston, MA (2005). https://doi.org/10.1007/0-387-26350-0_7

29. Winikoff, M., Dignum, V., Dignum, F.: Why bad coffee? Explaining agent plans with valuings. In: Gallina, B., Skavhaug, A., Schoitsch, E., Bitsch, F. (eds.) SAFECOMP 2018. LNCS, vol. 11094, pp. 521–534. Springer, Cham (2018). https://doi.org/10.1007/978-3-319-99229-7_47

BEN: An Agent Architecture
for Explainable and Expressive Behavior
in Social Simulation

Mathieu Bourgais[1]([✉]), Patrick Taillandier[2], and Laurent Vercouter[1]

[1] Normandie Univ, INSA Rouen Normandie, LITIS,
76000 Rouen, France
`mathieu.bourgais@insa-rouen.fr`
[2] MIAT, INRA, 31000 Toulouse, France

Abstract. Social Simulations are used to study complex systems featuring human actors. This means reproducing real-life situations involving people in order to explain an observed behavior. However, there are actually no agent architectures among the most popular platforms for agent-based simulation enabling to easily model human actors. This situation leads modelers to implement simple reactive behaviors while the EROS principle (Enhancing Realism Of Simulation) fosters the use of psychological and social theory to improve the credibility of such agents. This paper presents the BEN architecture (Behavior with Emotions and Norms) that uses cognitive, affective and social dimensions for the behavior of social agents. This agent architecture has been implemented in the GAMA platform so it may be used by a large audience to model agents with a high level explainable behavior. This architecture is used on an evacuation case, showing how it creates believable behaviors in a real case scenario.

Keywords: Social simulation · Agent architecture · Cognition · Emotions · Evacuation

1   Introduction

These last years, agent-based simulation has been used to study complex systems featuring human actors; the community is now speaking of social simulation [19]. The main goal is to reproduce real life situations involving hundreds or thousands of simulated humans in order to better understand interactions leading to an observed result.

To be as close as possible to case studied, social simulations have to integrate social agents with a behavior as close as possible to the human behavior. The creation of believable social agents implies the reproduction of complex processes simulating the human reasoning [44]. Such systems lead to a closer behavior to the one expected. However, the obtained behavior may be hard to explain and to express with high level concepts.

© Springer Nature Switzerland AG 2019
D. Calvaresi et al. (Eds.): EXTRAAMAS 2019, LNAI 11763, pp. 147–163, 2019.
https://doi.org/10.1007/978-3-030-30391-4_9

This is the meaning of the KISS (Keep It Simple, Stupid) principle [6] which invites modelers to keep a simple behavior model so it can be explainable by simple rules at any moment in the simulation. This principle has been discussed over the years, leading to the KIDS (Keep It Descriptive, Stupid) principle [18] that favors more descriptive models to gain realism and then the EROS (Enhancing Realism Of Simulation) principle [22] which calls for the use of cognitive, affective and social dimensions to improve the credibility of social agents. Therefore, the problem is to model realistic simulated humans with an explainable behavior at a high level.

To tackle this issue, this paper presents BEN (Behavior with Emotions and Norms), a modular agent architecture integrating cognition, emotions, emotional contagion, personality, norms and social relations. Each of these components relies on psychological or social theories, helping a modeler to improve the credibility of simulated humans and ensuring an explainable behavior with high level concepts [24].

This architecture is implemented and integrated within GAMA [48], a modeling and simulation platform aiming to be used by a large audience. The goal is to create a tool that may even be used by modelers who are not expert in programming, without loosing the expressivity for the behavior developed. This implementation is explained in this paper through the example case of the evacuation of a nightclub, showing it succeeds to handle a real-life scenario and still provide a behavior with a high degree of explainability.

This paper is structured as follows: Sect. 2 reviews existing works to create social agents with a cognitive behavior, an emotional engine or social relations but also the existing agent architecture in popular simulation platforms. In Sect. 3, a formalism is proposed to deal with the mental state of the agent in terms of cognition, emotion and social relations. Section 4 describes the BEN architecture which relies on the aforementioned formalism. Section 5 presents the implementation of BEN through an example to illustrate how it can be used on a model of evacuation to create an explainable and believable behavior for agents simulating humans. Finally, Sect. 6 serves as a conclusion.

2   Related Works

Creating a believable social agent with an explainable behavior may be complex [29]. To ease this process, simulation platforms and behavior architecture have been developed by the community. These existing works are presented in this section.

2.1   Frameworks and Platforms for Simulations

Among the various agent-based platforms [27], some like JACK [21] or Jadex [38] implement the BDI (Belief Desire Intention) [10] paradigm, giving a cognitive behavior to agents, based on modal logic [15]. The addition of cognition helps creating more believable agents [2] but these platforms are not suitable for thousands of agents required in simulation.

To overcome this problem, Sing and Padgham [42] propose to connect a simulation platform to an existing BDI framework (like JACK or Jadex) and, with the same idea [37], the Matsim platform [8] has been linked with the GORITE BDI framework [40]. These works require a high level in computer science, making it difficult to use by modelers with low level programming skills.

Frameworks like Repast [16] or MASON [31] are dedicated simulations tools which improve existing programming languages to ease the development of agent-based simulations. Agents are described by Java classes and the framework is used to describe the scheduling of the execution of all the classes and the output of the simulation. This means these softwares do not offer specific agent architecture to control or to explain the agents' behavior.

On the other hand, simulation platforms like Netlogo [50] or GAMA [48] are dedicated softwares with their own programming language, their own interface and their own interpreter and compiler. They are made to be easy to use by people with low level in programming skills and they can handle thousands of agents during simulations, making them usable for the definition of explainable models made by experts of the studied fields.

By default, these platform do not propose any particular architecture for the agent behavior: modelers have to define these behaviors with "if-then-else" rules. However, there exist plugins, for NetLogo [41] and for GAMA [47], to use agent architectures based on the BDI paradigm in order to create simulated humans with a more complex and more believable behavior. They both provide the agents with high level concepts such as beliefs and intentions and GAMA's plugin goes beyond, offering a reasoning engine, leading agents to make decisions based on the perception of its environment.

2.2 Agent Architectures for Social Simulations

Using behavioral architectures enables modelers to define more easily credible and explainable social agents as these architectures offer high level concepts from works in psychology and sociology for the decision making. Among the numerous agent architectures [7], some of the most known are presented in this section.

SOAR [28] and ACT-R [13] are two cognitive architectures, grounded on works from psychology. Agents have access to a long and a short term memory, making a decision based on the previous experiments in a given context. These approaches are more complex than the BDI paradigm, making the decision making process more credible. However, they require heavy computation time, which makes them less pertinent for social simulations involving thousand simulated actors.

CLARION [45] represents another proposition of cognitive architecture. The agent's reasoning is divided between four sub-systems, each one manipulating explicit and implicit elements to make a decision in a given context. To our knowledge, CLARION is still a theoretical architecture which has not been implemented in any simulation platform.

Another approach consists in building the reasoning engine around the emotions of the agents. For example, EMA [20] is based on the cognitive appraisal theory of emotion [4] developed by Smith and Lazarus [43] while DETT [49] is based on the OCC [36] theory of emotions. Both those systems creates emotions by assessing the perceptions of the environment and then infer a behavior from the emotional state of the agent.

eBDI [23] relies on OCC theory too but it also uses a BDI architecture to make decisions. This means emotions are created through perceptions and then act upon beliefs, desires and intentions. Finally, these modified mental states are used to make a decision. This proposition has not been yet integrated to a simulation platform.

Finally, some researchers propose to rely on the social context of the agent to describe its behavior: this is done with normative architectures. EMIL-A [3] and NoA [26] describe the agent's behavior with social norms, obligations and sanctions. In other words, an agent makes a decision depending on the state of the normative system at the level of a society of agents.

BOID [12] and BRIDGE [17] propose to combine a normative architecture with a BDI paradigm, leading the agent to take into account the social system when making a decision. However, contrary to EMIL-A and NoA, the agent has personal beliefs, desires and intention, creating a more heterogeneous and credible behavior. But, to our knowledge, these architectures have not been implemented in simulation platform in order to deal with thousand of simulated actors.

2.3   Synthesis

To comply with the EROS principle, modelers need architectures proposing as much psychological and social dimensions as possible. Currently, as shown in this section, there does not exists a single architecture proposing at the same time, cognition, affective dimensions and social dimension for simulation. The only attempts, to our knowledge, to combine more than two traits have used the notion of personality, to combine cognition with emotions and emotional contagion [30] or to combine cognition with emotions and social relations [35].

In this paper, we tackle this issue by proposing BEN (Behavior with Emotions and Norms), an agent architecture featuring cognition, emotions, personality, emotional contagion, social relations and norm management. To implement it, we have based our work on the existing cognitive architecture provided by GAMA. To ease the use of BEN, we have implemented it using the principles of GAMA that has proved its ease of use [32,39] thanks to its modeling language GAML that we extended.

3   Formalization of Mental States

With the BEN architecture, an agent manipulates cognitive mental states, emotions and social relations to make a decision. These notions, and the formalism used to represent them, are presented in this section.

3.1 Representing the World with Predicates

Predicates are used to unify the representation of the information from the world from the agent's point of view. $\mathbf{P}_j(\mathbf{V})$ represents a general predicate with the following elements:

- **P:** the identifier of the predicate.
- **j:** the agent causing the information.
- **V:** a set of values stored by the predicate.

Depending on the context, this general representation may change. $\mathbf{P}_j(\mathbf{V})$ represents an information with no particular value attached, $\mathbf{P}(\mathbf{V})$ represents an information caused by no particular agent and $\mathbf{P}$ stands for an information with no particular value, caused by no particular agent.

For example, the information there is a fire in the environment is represented by the predicate **fire**. If this fire is caused by agent *Bob*, it is represented by **fire**$_{Bob}$. Finally, if this fire caused by Bob is at a location (x;y), this information is represented by **fire**$_{Bob}$(**location** :: $(\mathbf{x}, \mathbf{y})$).

3.2 Reasoning According to Cognitive Mental States

With BEN, an agent has cognitive mental states representing its thoughts on the world. $\mathbf{M}_i(\mathbf{PMEm}, \mathbf{Val}, \mathbf{Li})$ represents a general cognitive mental state possessed by agent i with the following elements:

- **M:** the modality, indicating the type of the cognitive mental state (e.g. a belief).
- **PMEm:** the object of the cognitive mental state. It could be a predicate (P), another cognitive mental state (M) or an emotion (Em).
- **Val:** a real value which meaning depends on the modality. It enables to compare two cognitive mental states with the same modality and the same object.
- **Li:** a lifetime value indicating how long the cognitive mental state stays in the agent's knowledge.

In BEN, cognition is based on the BDI model [10], stating an agent needs beliefs, desires and intentions. Also, to link cognition with affective and social dimensions, BEN features 6 different modalities:

- **Belief:** represents what the agent beliefs about the world. The meaning of the attached value is the strength given by the agent to the belief.
- **Uncertainty:** represents an uncertain information about the world. The meaning of the attached value is the importance given to that uncertainty by the agent.
- **Desire:** represents a state of the world the agent want to reach. The meaning of the attached value is the priority of this desire, compared to other desires.
- **Intention:** represent a state of the world the agent is willing to achieve. The meaning of the attached value is the priority of this intention, compared to other intentions.

- **Ideal:** represents an information which is socially appraised by the agent. The attached value is a praiseworthiness value given by the agent.
- **Obligation:** represents a state of the world the agent ought to reach. The meaning of the attached value is the priority of this obligation, compared to other obligations.

3.3   Representing Emotions and Social Relations

Emotions in BEN are based on the OCC theory [36] which means emotions are valued answers to the appraisal of a situation. $Em_i(P, Ag, I, De)$ represent an emotion possessed by agent i with the following elements:

- **Em:** the name of the emotion.
- **P:** the predicate about which the emotion is felt.
- **Ag:** the agent responsible for the emotion.
- **I:** the intensity, positive or null, of the emotion.
- **De:** the decay value for the emotion's intensity.

This representation enables the agent to have multiple emotions at the same time, all on different predicates. Also, this representation can be adapted, with $Em_i(P, Ag)$ representing an emotion with no particular intensity nor decay value.

Finally, each agent may store social relations with other agents simulating human actors. These relations are based on the work of Svennevig [46] who identifies four minimal dimensions to describe a social relation between two people. In BEN, $R_{i,j}(L, D, S, F, T)$ represents a social relation, from agent i towards agent j with the following elements:

- **R:** the identifier of the social relation.
- **L:** a real value between -1 and 1 standing for the degree of liking. A value of -1 indicates agent j is hated, a value of 1 indicates agent j is liked.
- **D:** a real value between -1 and 1 standing for the degree of dominance. A value of -1 indicates agent j is dominating, a value of 1 indicates agent j is dominated.
- **S:** a real value between 0 and 1 standing for the degree of solidarity. A value of 0 indicates no solidarity with agent j, a value of 1 indicates a complete solidarity with agent j.
- **F:** a real value between 0 and 1 standing for the degree of familiarity. A value of 0 indicates no familiarity with agent j, a value of 1 indicates a complete familiarity with agent j.
- **T:** a real value between -1 and 1 standing for the degree of trust. A value of -1 indicates mistrust against agent j, a value of 1 indicates a complete trust towards agent j.

With this definition, social relations do not have to be symmetric; between two agents i and j, $R_{i,j}(L,D,S,F,T)$ is not obviously equal to $R_{j,i}(L,D,S,F,T)$.

4   An Agent Architecture with Cognitive, Affective and Social Dimensions

The BEN architecture represents the main contribution of this article. In this section, we explain how an agent using BEN makes a decision with cognition, emotions, emotional contagion, personality, social relations and norms. With these dimensions, an agent simulating an actor may react to a change in the environment and still explain its behavior with high level concepts.

4.1   Global Presentation of the Architecture

Figure 1 represents the theoretical BEN architecture, providing cognitive, affective and social dimensions to agents simulating human actors. It is made up of four modules, each composed of several processes, communicating with the agent's knowledge, all of this seating on the agent's personality.

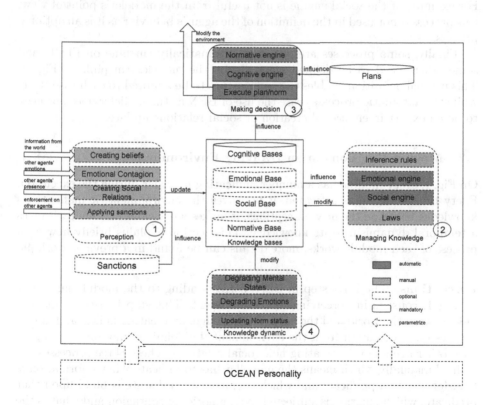

Fig. 1. The BEN architecture (Color figure)

The personality component is based on the OCEAN [33] theory which defines fives parameters (Openness, Consciousness, Extroversion, Agreeableness, Neurotism) that are sufficient to represent a personality. To ease the use of BEN, these personality traits are the only parameter a modeler may access; they are used to compute all the other parameters needed by the various processes: probability to remove a plan or an intention unfulfilled in the cognitive part, charisma and emotional receptivity for the emotional contagion, initial intensity and decay for emotions created by the engine, update values of social relations obtained with the social engine, and obedience for the normative engine.

The agent's knowledge is composed by cognitive bases, containing cognitive mental states as formalized in section, an emotional base, a social base and a base of norms. This knowledge may evolve through the simulation, which is not the case of plans for the cognitive engine and sanctions for the normative engine, which are stored in dedicated bases, out of the agent's knowledge, as seen on Fig. 1.

Each module, and each process of each module, may be mandatory (in plain line on Fig. 1) or optional (in dash line on Fig. 1); the optional modules and processes may be deactivated by modelers if not necessary in the case studied. For example, if the social engine is not useful from the modeler's point of view, this process is not used in the definition of the agent's behavior as it is an optional process.

Finally, some processes are executed automatically (in blue on Fig. 1) and some others need to be defined manually by the modeler (in pink on Fig. 1). This manual definition enables the architecture to be adapted to each case study while the automatic processes ease the use of BEN as the modeler does not need to be an expert in emotional creation or social relations update.

4.2   Making Decisions in an Evolving Environment

On Fig. 1, every module has a number, indicating its order during the execution. Every time an agent is activated, it perceives the environment, it manages its knowledge based on the new perceptions, it makes a decisions and finally it gives a temporal dynamism to its knowledge. This section explains briefly how each process of each module works, more details can be found in a previous work [9].

Perceptions. The First step in BEN, corresponding to the module number 1 on Fig. 1, consists in perceiving the environment. This step is used to make a link between the world and the agent's knowledge, by creating beliefs and uncertainties on information from the environment, by defining emotional contagion with other agents or by creating new social relations. These three processes are defined manually which means the modeller has to indicate what information is transformed as a predicate and which cognitive mental state is build upon that predicate, which emotion is subjected to an emotional contagion and what is the initial value for each dimension of a new social relation. The last process of this module enables an agent to execute sanctions during the enforcement done on the other agents perceived.

Adding a belief is an important process in BEN as it triggers different rules. Precisely, adding a belief $Belief_A(X)$:

- removes belief $Belief_A(notX)$.
- removes intention $Intention_A(X)$.
- removes desire $Desire_A(X)$ if intention $Intention_A(X)$ has just been removed.
- removes uncertainty $Uncertainty_A(X)$ or $Uncertainty_A(notX)$.
- removes obligation $Obligation_A(X)$.

With the same principle, adding uncertainty $Uncertainty_A(X)$:

- removes uncertainty $Uncertainty_A(notX)$.
- removes belief $Belief_A(X)$ or $Belief_A(notX)$

All these processes are defined inside a perception, which may be parameterized. A modeller indicates a distance of perception or a geometry inside which the perception is done, but also specifies which agents are perceived. As it is, this module and all its processes may adapt to any case study in social simulation.

Managing Knowledge. The second step of BEN, corresponding to the module number 2 on Fig. 1, enables the agent to manage its knowledge after the perception and before making decision. In this phase, modelers may define inference rules, which enable to create or remove any cognitive mental state depending on the actual status of the agent's knowledge. For example, a modeler may define that an agent has a new desire $Desire_A(Y)$ if this agent has a belief $Belief_A(X)$. On the same model, laws may be defined to create obligations if the obedience value of the agent, computed from its personality, is great enough.

During this second step, an emotional engine creates emotions based on the agent's knowledge. This process is done according to the OCC theory [36] and its formalism with the BDI model [1]. For example, an emotion of joy about a predicate P is created according to the following rule: $Joy_i(P_j,j) \overset{\text{def}}{=} Belief_i(P_j)$ & $Desire_i(P)$. The complete process and all the rules to create 20 emotions automatically, with no intervention from the modeler, are detailed in a previous work [9].

Finally, a social engine may be executed during this second step of the architecture. It updates the social relation with the other agents perceived, depending on the knowledge previously acquired. All this process and the equations to compute automatically the new value for each dimension of a social relation are explained in a previous work [9].

Making Decision. The third step of BEN, corresponding to the module number 3 on Fig. 1, is the only mandatory part. This module enables the agent to make decisions and then execute an action, all of this through a cognitive engine over which a normative engine may be added. It is executed automatically, with no need of intervention from the modeler.

The cognitive architecture is based on the BDI model [10]: the agent has intentions based on its desires and one of the intentions as a current intention. The modeler defines plans of action that indicates what action an agent has to do for a particular current intention in a given context; the plan chosen is kept as the current plan. The normative engine works the same way as the cognitive engine, with obligations as desires and norms as plans. The only difference is an obedience value that can be added to norms and obligations. More details may be found in previous works [9].

Temporal Dynamics. The final part of the architecture, corresponding to the module number 4 on Fig. 1, gives a temporal dynamic to the agent's behavior. This is done automatically by degrading the cognitive mental states and the emotions and by updating the status of each norm.

The degradation of cognitive mental states decreases the lifetime of each cognitive mental state stored by the agent. This mechanism enables an agent to forget, after a certain time, a belief, a desire, etc. The degradation of emotions consists in subtracting the decay value of each emotion to each intensity. With this process, an emotion fades away, unless it is created again, for example with the emotional engine or through the emotional contagion process.

Finally, the last process updates the status of each norm, indicating if it was usable in the current context, and in the case it was usable, has it been used or not by the agent. This system enables to ease the enforcement of norms in a later perception, as each norm indicates its status instead of computing this status in another context.

5   Simulating the Evacuation of a Nightclub

The architecture defined in Sect. 4 has been implemented in the modeling and simulation platform GAMA [48], extending the GAML programming language to help modelers define social agents with cognitive, affective and social dimensions to express their behavior. This implementation is used on the example case of an evacuation of a nightclub in fire, as detailed in this section.

5.1   Presentation of the Example Case

The 27th of January 2013, the Kiss Nightclub in Santa Maria, Rio Grande do Sul state is Brazil, was set in fire at the end of a show lead by a local music band. The ceiling caught fire because of fireworks, emitting toxic smokes which lead to the death of 242 people. The official investigation put light on various factors which aggravated the tragedy: there were between 1200 and 1400 people in the building that could normally handle 641 people, there was only one entrance/exit door, there was no smoke detector nor alarms and finally, the exit signs were showing the direction of the restrooms. The investigation also shows that most of the deceases were due to asphyxia, near the restrooms [5].

Our goal here is to reproduce the behavior of people caught in this tragedy in the most credible way possible. In other words, we are using the BEN architecture to create the agents' behaviors in order to get a result as close as possible as what happened in this nightclub during the fire.

5.2   Modeling the Behavior of Human Actors with BEN

The initial agent's knowledge, at the start of the simulation, can be divided into three types: beliefs about the world, initial desires and social relations with friends. Also, each agent has a personality. Table 1 indicates how few of these initial knowledge are formalized with BEN.

Table 1. Example of agent's initial knowledge

Statement	Formalisation	Description
A belief on the exact position of the exit door	$Belief_i(exitDoor, lifetime1)$	Each agent has a belief about the precise location of the exit door with a lifetime value at $lifetime1$
A desire there is no fire	$Desire_i(notFire, 1.0)$	Each agent wish there is no fire in the nightclub with a priority of 1.0. This desire cannot lead to an action (no action plan are defined to answer it)
A relation of friendship with another agent	$R_{i,j}(L, D, S, F, T)$	Each agent i is likely to have a social relation with agent j, representing its friend

The first step of BEN is the perception of the environment. We need to define what an agent perceives and how it affects its knowledge. Here are examples of the agent's perceptions:

- Perceiving the exit door updates the beliefs related to it.
- Perceiving the fire adds the belief there is a fire.
- Perceiving the smoke adds the belief about the level of smoke perceived.
- Perceiving other agents enables to create social relations with them. An emotional contagion about the fear of a fire is also defined.

Once the agent is up to date with its environment, its overall knowledge has to adapt to what it has perceived. This is done with the definition of inference rules and laws:

- A law creates the obligation to follow the exit signs if there is a reasonable doubt (modeled by the obedience value attached to the law and the quantity of smoke perceived) of a catastrophe.
- An inference rule adds the desire to flee if the agent has a belief there is fire.
- An inference rule adds an uncertainty there is a fire if the agent has a belief there is smoke.
- An inference rule adds the desire to flee if the agent has a fear emotion about the fire with an intensity greater than a given threshold.

With the execution of inference rules and laws, each agent creates emotions with the emotional engine. In this case, the presence of an uncertainty about the fire (added through the inference rule concerning the belief about smoke) with the initial desire that there is no fire produces an emotion of fear, which intensity is computed depending on the quantity of smoke perceived.

Once the agent has the desire to flee (because it perceived the fire or its fear of a fire had an intensity great enough), it needs action plans and norms to indicate what it has to do. Table 2 shows the definition of some action plans and norms used by the agent to answer its intention to flee, depending of the context it perceives.

Table 2. Action plans and norms answering the fleeing intention

Conditions	Actions	Commentaries
The agent has a good visibility and has a belief on the exact location of the exit door	The agent runs to the exit door	In this plan, the agent runs to the exit door following the shortest path
The agent has a good visibility and has no belief about the location of the door	The agent follows the agent in its field of view with the highest trust value among its social relations	This norm works with the trust value of social relations created during the simulation
The agent has a bad visibility and has the obligation to follow signs	The agent goes to the restrooms	In this norm, the agent comply with the law that indicates to follow exit signs
The agent has a bad visibility and has a belief exit signs are wrong	The agent moves randomly	In this plan, the agent moves randomly in the smoke

The social relation defined with a friend may also be used to define plans to help one's friend if it is lost in smoke. This plan consists in finding the friend and telling him the location of the exit door.

As the situation evolves during the simulation, an agent may change its current plan. For example, if an agent leaves the smoke area while fleeing to the restroom, it may perceive the exit, and go there instead of following the signs.

The complete model can be found at this address: https://github.com/mathieuBourgais/ExempleThese.

5.3   Results and Discussion

At the start of the simulation, agents are placed randomly in the recreated Kiss Nightclub with a personality initialised by a Gaussian distribution centered on 0.5 and with a standard deviation of 0.12 for each dimension. The spread of the smoke is modeled according to an official report from the french government [14]; an agent is considered dead after 50 s in the heavy smoke.

Figure 2 shows a visual result of the simulation where the black lines represent the walls of the nightclub, the grey squares represent the smoke and the triangles represent the simulated actors. The color of each triangle indicates the plan followed. A video of the simulation can be found on the following address: https://github.com/mathieuBourgais/ExempleThese.

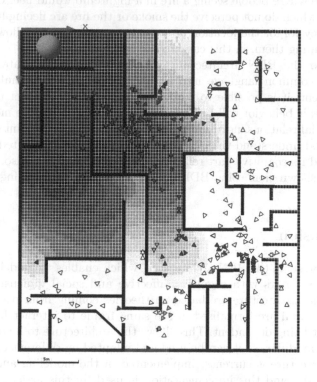

Fig. 2. Simulation of the Kiss Nightclub's evacuation

As the exact number of people in the nightclub is not known, we tested three cases: 1200 people at the beginning, 1300 people at the beginning and 1400 people at the beginning. The statistical results obtained in Table 3 are computed from 10 simulations for each scenario.

Table 3. Number of agents dead in the simulation of the Kiss Nightclub fire

Number of agents	1200	1300	1400
Mean value	230.2	237.7	249.4
Standard deviation	20.1	15.6	32.6

Statistical results indicate our model is well calibrated to reproduce the real life case where 242 people died. However, the main result concerns the explainability and the expressivity of the model. The video of the simulation shows various behavior patterns which may be expressed with high level concepts thanks to BEN.

For example, a lot of agents leave the club at the beginning of the simulation because they directly perceived the fire. This behavior seems corresponding to a real life case where people seeing a fire in a nightclub would flee. On the other hand, agents which do not perceive the smoke or the fire are fleeing later. During that time, they forgot the location of the exit so they had to follow the official exit signs, leading them, in this case, to the restrooms.

Thanks to the BEN architecture, we were able to translate a behavior expressed in common language into an actual behavior for simulated actors. At any moment, it is possible to pause the simulation to inspect the behavior of an agent; this behavior will be expressed in terms of cognitive mental states, emotions, social relations, norms and plans which is, from our point of view, easier to read and understand than equations. This point is supported by the fact that BEN and its cognitive part relies on folk psychology [34]. Also, some works [11,25] have shown that using BDI and emotions helps explaining the agents' behavior.

6   Conclusion

This article presents the BEN architecture, which enables to model agents simulating human actors with cognitive, affective and social dimensions. All the features of the architecture are based on theories coming from psychology and social sciences and are formalised in the same frame to interact between each other without being dependent. This allows the architecture to be domain independent and modular, so it can be used and adapted on different contexts.

This architecture is currently implemented in the modeling and simulation platform GAMA and this implementation is used, in this article, on the case study of the evacuation of a nightclub in fire. This example shows BEN achieves to produce a complex and credible behavior but maintaining its high level explainability.

Acknowledgments. This work is partially supported by two public grants overseen by the French National Research Agency (ANR) as part of the program PRC (reference: ESCAPE ANR-16-CE39-0011-01 and ACTEUR ANR-14-CE22-0002).

References

1. Adam, C.: Emotions: from psychological theories to logical formalization and implementation in a BDI agent. Ph.D., thesis, INP Toulouse (2007)
2. Adam, C., Gaudou, B.: BDI agents in social simulations: a survey. Knowl. Eng. Rev. **31**, 207–238 (2016)

3. Andrighetto, G., Conte, R., Turrini, P., Paolucci, M.: Emergence in the loop: simulating the two way dynamics of norm innovation. In: Dagstuhl Seminar Proceedings. Schloss Dagstuhl-Leibniz-Zentrum für Informatik (2007)
4. Arnold, M.B.: Emotion and Personality. Columbia University Press, New York (1960)
5. Atiyeh, B.: Brazilian kiss nightclub disaster. Ann. Burns Fire Disasters. **26**, 3 (2013)
6. Axelrod, R.: Advancing the art of simulation in the social sciences. In: Conte, R., Hegselmann, R., Terna, P. (eds.) Simulating social phenomena. LNE, vol. 456, pp. 21–40. Springer, Berlin (1997). https://doi.org/10.1007/978-3-662-03366-1_2
7. Balke, T., Gilbert, N.: How do agents make decisions? A survey. J. Artif. Soc. Soc. Simul. **17**(4), 13 (2014)
8. Balmer, M., et al.: MATSim-T: architecture and simulation times. In: Multi-agent Systems for Traffic and Transportation Engineering (2009)
9. Bourgais, M.: Towards cognitive, affective and social agents in the simulation. Theses, Normandie Université (2018)
10. Bratman, M.: Intentions, Plans, and Practical Reason. Harvard University Press, Cambridge (1987)
11. Broekens, J., Harbers, M., Hindriks, K., van den Bosch, K., Jonker, C., Meyer, J.-J.: Do you get it? User-evaluated explainable BDI agents. In: Dix, J., Witteveen, C. (eds.) MATES 2010. LNCS (LNAI), vol. 6251, pp. 28–39. Springer, Heidelberg (2010). https://doi.org/10.1007/978-3-642-16178-0_5
12. Broersen, J., Dastani, M., Hulstijn, J., Huang, Z., van der Torre, L.: The BOID architecture: conflicts between beliefs, obligations, intentions and desires. In: Proceedings of the Fifth International Conference on Autonomous Agents, pp. 9–16. ACM (2001)
13. Byrne, M.D., Anderson, J.R.: Perception and action. In: The Atomic Components of Thought (1998)
14. Chivas, C., Cescon, J.: Formalisation du savoir et des outils dans le domaine des risques majeurs (dra-35) - toxicité et dispersion des fumées d'incendie phénoménologie et modélisation des effets. Technical report, INERIS (2005)
15. Cohen, P.R., Levesque, H.J.: Intention is choice with commitment. Artif. Intell. **42**(2–3), 213–261 (1990)
16. Collier, N.: RePast: an extensible framework for agent simulation. Univ. Chicago's Soc. Sci. Res. **36**, 2003 (2003)
17. Dignum, F., Dignum, V., Jonker, C.M.: Towards agents for policy making. In: David, N., Sichman, J.S. (eds.) MABS 2008. LNCS (LNAI), vol. 5269, pp. 141–153. Springer, Heidelberg (2009). https://doi.org/10.1007/978-3-642-01991-3_11
18. Edmonds, B., Moss, S.: From KISS to KIDS – An 'Anti-simplistic' modelling approach. In: Davidsson, P., Logan, B., Takadama, K. (eds.) MABS 2004. LNCS (LNAI), vol. 3415, pp. 130–144. Springer, Heidelberg (2005). https://doi.org/10.1007/978-3-540-32243-6_11
19. Gilbert, N., Troitzsch, K.: Simulation for the Social Scientist. McGraw-Hill Education (UK), New York (2005)
20. Gratch, J., Marsella, S.: A domain-independent framework for modeling emotion. Cogn. Syst. Res. **5**, 269–306 (2004)
21. Howden, N., Rönnquist, R., Hodgson, A., Lucas, A.: Jack intelligent agents-summary of an agent infrastructure. In: 5th International Conference on Autonomous Agents (2001)

22. Jager, W.: Enhancing the realism of simulation (EROS): on implementing and developing psychological theory in social simulation. J. Artif. Soc. Soc. Simul. **20**(3), 14 (2017)
23. Jiang, H., Vidal, J.M., Huhns, M.N.: EBDI: an architecture for emotional agents. In: Proceedings of the 6th International Joint Conference on Autonomous Agents and Multiagent Systems. ACM (2007)
24. Kaptein, F., Broekens, J., Hindriks, K., Neerincx, M.: Self-explanations of a cognitive agent by citing goals and emotions, October 2017
25. Kaptein, F., Broekens, J., Hindriks, K., Neerincx, M.: The role of emotion in self-explanations by cognitive agents. In: 2017 Seventh International Conference on Affective Computing and Intelligent Interaction Workshops and Demos (ACIIW), pp. 88–93. IEEE (2017)
26. Kollingbaum, M.J.: Norm-governed practical reasoning agents. Ph.D., thesis, University of Aberdeen Aberdeen (2005)
27. Kravari, K., Bassiliades, N.: A survey of agent platforms. J. Artif. Soc. Soc. Simul. **18**(1), 11 (2015)
28. Laird, J.E., Newell, A., Rosenbloom, P.S.: Soar: an architecture for general intelligence. Artif. Intell. **33**, 1–64 (1987)
29. Langley, P., Meadows, B., Sridharan, M., Choi, D.: Explainable agency for intelligent autonomous systems. In: Twenty-Ninth IAAI Conference (2017)
30. Lhommet, M., Lourdeaux, D., Barthès, J.-P.: Never alone in the crowd: a microscopic crowd model based on emotional contagion. In: Web Intelligence and Intelligent Agent Technology (WI-IAT). IEEE (2011)
31. Luke, S., Cioffi-Revilla, C., Panait, L., Sullivan, K., Balan, G.: Mason: a multiagent simulation environment. Simulation **81**, 517–527 (2005)
32. Macatulad, E.G., Blanco, A.C.: 3DGIS-based multi-agent geosimulation and visualization of building evacuation using GAMA platform. Int. Arch. Photogrammetry Remote Sens. Spatial Inf. Sci. **40**, 87 (2014)
33. McCrae, R.R., John, O.P.: An introduction to the five-factor model and its applications. J. Pers. **60**(2), 175–215 (1992)
34. Norling, E.: Folk psychology for human modelling: extending the BDI paradigm. In: Proceedings of the Third International Joint Conference on Autonomous Agents and Multiagent Systems-Volume 1, pp. 202–209. IEEE Computer Society (2004)
35. Ochs, M., Sabouret, N., Corruble, V.: Simulation of the dynamics of nonplayer characters' emotions and social relations in games. IEEE Trans. Comput. Intell. AI Games **1**(4), 281–297 (2009)
36. Ortony, A., Clore, G.L., Collins, A.: The Cognitive Structure of Emotions. Cambridge University Press, Cambridge (1990)
37. Padgham, L., Nagel, K., Singh, D., Chen, Q.: Integrating BDI agents into a MATSim simulation. In: Proceedings of the Twenty-First European Conference on Artificial Intelligence, pp. 681–686. IOS Press (2014)
38. Pokahr, A., Braubach, L., Lamersdorf, W.: Jadex: a BDI reasoning engine. In: Bordini, R.H., Dastani, M., Dix, J., El Fallah Seghrouchni, A. (eds.) Multi-Agent Programming. MSASSO, vol. 15, pp. 149–174. Springer, Boston (2005). https://doi.org/10.1007/0-387-26350-0_6
39. Ramli, N.R., Razali, S., Osman, M.: An overview of simulation software for non-experts to perform multi-robot experiments. In: ISAMSR. IEEE (2015)
40. Rönnquist, R.: The goal oriented teams (GORITE) framework. In: Dastani, M., El Fallah Seghrouchni, A., Ricci, A., Winikoff, M. (eds.) ProMAS 2007. LNCS (LNAI), vol. 4908, pp. 27–41. Springer, Heidelberg (2008). https://doi.org/10.1007/978-3-540-79043-3_2

41. Sakellariou, I., Kefalas, P., Stamatopoulou, I.: Enhancing NetLogo to aimulate BDI communicating agents. In: Darzentas, J., Vouros, G.A., Vosinakis, S., Arnellos, A. (eds.) SETN 2008. LNCS (LNAI), vol. 5138, pp. 263–275. Springer, Heidelberg (2008). https://doi.org/10.1007/978-3-540-87881-0_24

42. Singh, D., Padgham, L.: OpenSim: a framework for integrating agent-based models and simulation components. In: Frontiers in Artificial Intelligence and Applications-Volume 263: ECAI 2014. IOS Press (2014)

43. Smith, C.A., Lazarus, R.S., et al.: Emotion and adaptation. In: Handbook of Personality: Theory and Research, pp. 609–637 (1990)

44. Sun, R.: Cognition and Multi-agent Interaction: From Cognitive Modeling to Social Simulation. Cambridge University Press, Cambridge (2006)

45. Sun, R.: The importance of cognitive architectures: an analysis based on clarion. J. Exp. Theor. Artif. Intell. **19**(2), 159–193 (2007)

46. Svennevig, J.: Getting Acquainted in Conversation: A Study of Initial Interactions. John Benjamins Publishing, Amsterdam (2000)

47. Taillandier, P., Bourgais, M., Caillou, P., Adam, C., Gaudou, B.: A BDI agent architecture for the GAMA modeling and simulation platform. In: Nardin, L.G., Antunes, L. (eds.) MABS 2016. LNCS (LNAI), vol. 10399, pp. 3–23. Springer, Cham (2017). https://doi.org/10.1007/978-3-319-67477-3_1

48. Taillandier, P., et al.: Building, composing and experimenting complex spatial models with the GAMA platform. GeoInformatica **23**, 299–322 (2018)

49. Van Dyke Parunak, H., Bisson, R., Brueckner, S., Matthews, R., Sauter, J.: A model of emotions for situated agents. In: Proceedings of the Fifth International Joint Conference on Autonomous Agents and Multiagent Systems. ACM (2006)

50. Wilensky, U., Evanston, I.: NetLogo: Center for connected learning and computer-based modeling, Northwestern University, Evanston, IL (1999)

Planning and Argumentation

Temporal Multiagent Plan Execution: Explaining What Happened

Gianluca Torta[(✉)] [ID], Roberto Micalizio [ID], and Samuele Sormano

Dipartimento di Informatica, Università di Torino, Torino, Italy
{gianluca.torta,roberto.micalizio}@unito.it,
samuele.sormano@edu.unito.it

Abstract. The paper addresses the problem of explaining failures that happened during the execution of Temporal Multiagent Plans (TMAPs), namely MAPs that contain both logic and temporal constraints about the action conditions and effects. We focus particularly on computing explanations that help the user figure out how failures in the execution of one or more actions propagated to later actions. To this end, we define a model that enriches knowledge about the nominal execution of the actions with knowledge about (faulty) execution modes. We present an algorithm for computing diagnoses of TMAPs execution failures, where each diagnosis identifies the actions that executed in a faulty mode, and those that failed instead because of the propagation of other failures. Diagnoses are then integrated with temporal explanations, that detail what happened during the plan execution by specifying temporal relations between the relevant events.

Keywords: Temporal Multiagent Plans · Model-based diagnosis · SMT

1 Introduction

Multiagent plans (MAPs) are an efficient way for accomplishing complex goals. The underlying principle is to decompose a given complex goal into subgoals, and then organize the activities of a team of agents so as that each agent achieves a subgoal autonomously while coordinating with others. Plan execution, however, is not always straightforward. The actual execution of actions, in fact, can be affected by failures. When a failure occurs, detecting and diagnosing it is of primary importance in order to resume the nominal execution. As pointed out in [2], in fact, when the behavior of a system is not explained, a human user makes up her own explanation, that not necessarily reflects the internal stance of the system.

The diagnosis of the execution of a multiagent plan (MAP) has been addressed in a number of works (see e.g., [9–11]), proposing different notions of plan diagnosis and different diagnostic methodologies. These works, however,

© Springer Nature Switzerland AG 2019
D. Calvaresi et al. (Eds.): EXTRAAMAS 2019, LNAI 11763, pp. 167–185, 2019.
https://doi.org/10.1007/978-3-030-30391-4_10

do not model time explicitly, but only implicitly by assuming a sequence of identical time steps at which atomic actions are performed. In these approaches, thus, it is not possible to model *durative actions* [5]; however, in real world scenarios, action duration (either within a nominal range, or with unexpected delays) can strongly affect the success of the agent's plan and of its interactions with others.

Other works have addressed the diagnosis of delayed actions in MAPs [12,13, 16]. Their objective is to provide the user with explanations of failures consisting only of actions delays; whereas, the logical effects of action failures (i.e., missing logical values that should hold as a consequence of an action) are not taken into consideration. This restriction limits the applicability of the methods by hindering their ability to handle cases of fault propagation from an action to another one due to a missing effect.

This paper, which significantly expands our previous work [17], contributes with a comprehensive framework addressing the diagnosis of a MAP execution by taking into account both missing effects and temporal deviations. We adopt a consistency-based notion of diagnosis [15]: a MAP diagnosis is a subset of actions whose non-nominal behavior is consistent with the observations received so far. We then argue that, in a setting with agents interactions and durative actions, such diagnoses may not be informative enough for helping a human user figure out what happened during the plan execution. As a remedy, we enrich diagnoses with *temporal explanations* that clarify how primary action failures may have affected other actions in the MAP, even those assigned to different agents. Figure 1 outlines the loop of inferences we aim at. This loop is substantially grounded on a consistency-based notion of diagnosis. Intuitively, a MAP P is properly encoded into a model $S_{P,\mathcal{H}}$ which takes into account an initial hypothesis $\mathcal{H}$ of nominal behavior. The model is therefore used for detecting discrepancies between the expected behavior of the whole P with the available observations about the actual execution of the plan. In case a discrepancy is found, a diagnosis task is activated with the aim of detecting a number of alternative consistency-based diagnoses. To increase the informative power of such diagnoses, they are complemented with a set of temporal explanations. Finally, a human supervisor has the chance to evaluate the alternative explanations and select the one to be chosen as the new current hypothesis $\mathcal{H}$. Although this presentation and selection phase is not addressed in this paper (as we underline in the picture by using dashed lines), we deem that the synthesis of such temporal explanations is a fundamental step to increase users' awareness.

Specifically, we propose a methodology to solve a diagnostic problem by inferring the set of all the preferred diagnoses with *minimal rank* [6], i.e., with the highest (order-of-magnitude) likelihood. Our approach is based on a single, centralized diagnostic reasoner that must diagnose the behavior of a multiagent system. Since we deal with both logic and temporal constraints to model faulty action modes, the computation of all the preferred diagnoses is made by exploiting a Satisfiability Modulo Theories (SMT) solver, that is able to handle both kinds of conditions. We model propagation by considering literals that are shared among the actions (i.e., produced as an effect by an action, and consumed as a

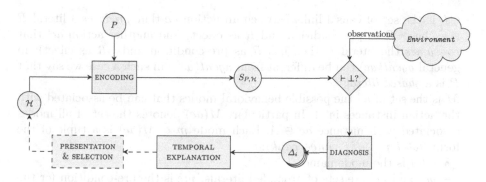

Fig. 1. The outline of the proposed TMAP diagnosis loop.

precondition by another action, even of a different agent). These shared literals can be considered as resources that are dynamically generated, and consumed, during the execution. To explain an action failure as an indirect consequence of a previous failure, thus, we focus on the events that affect the values of the literals shared by the two actions.

To the best of our knowledge, our proposal is the first one dealing with both temporal and logic aspects in the diagnosis of multiagent plans. The most similar work we are aware of is [4], where, however, the authors consider only plans with a limited number of discrete time steps amenable to a SAT encoding, and concentrate on conflicts among agents in the use of resources (e.g., road intersections).

The paper is organized as follows. In the next section we formalize the notion of Temporal Multi-Agent Plans (TMAPs). In Sect. 3 we introduce the Plan Execution Failure (PEF) diagnostic problem and the notion of preferred diagnosis, and in Sect. 4 we motivate and formally define the (temporal) explanations of diagnoses. In Sects. 5 and 6 we first describe how the relevant information of a PEF problem can be encoded in the input language of a SMT solver, and then we discuss how the PEF problem can be solved with a conflict-based search algorithm, and how explanations of a diagnosis can be computed. In Sect. 7, before conclusions, we discuss the experimental results we have obtained with an implementation of the proposed approach.

2   Temporal Multiagent Plans

We formalize a Temporal Multiagent Plan (TMAP) P as $\langle T, A, O, CL, M \rangle$:

- T is the team of cooperating agents $ag_1, ag_2, \ldots$
- A is the set of action instances $ac_1, ac_2, \ldots$ included in the plan, each of which is assigned to a specific agent $agent(ac_i)$;
- O is a set of order constraints, that specifies a total order relation over the actions of each agent $ag \in T$; each pair $\langle ac, ac' \rangle \in O$, $ac, ac' \in A$, $agent(ac) = agent(ac')$ means that ac is the *predecessor* of ac', and ac' is the *successor* of ac;

- CL is the set of causal links between an action ac that *produces* a literal R (i.e., has $\neg R$ as pre-condition and R as effect) and another action ac' that *consumes* the literal R (i.e., has R as pre-condition and $\neg R$ as effect); in general $agent(ac)$ can be different from $agent(ac')$; in such a case we say that R is a *shared literal*;
- M is the set of all the possible behavioral modes that can be associated with the action instances in A. In particular, $M(ac)$ denotes the set of all modes associated with instance $ac \in A$. Each mode $m \in M(ac)$ is a tuple of the form $\langle label, pre, eff, range, rank \rangle$:
  - *label* is the mode name;
  - *pre* and *eff* are sets of grounded literals: *pre* is the pre-condition for the execution of ac in mode m; whereas, *eff* is the set of effects obtained by performing ac in mode m^1;
  - *range* is an interval of time corresponding to the possible durations of the action when it behaves in mode m;
  - *rank* is a non-negative integer value representing the order-of-magnitude probability of the mode [6]: lower ranks correspond to higher probabilities.
  Set $M(ac)$ must contain at least one distinguished mode N (nominal) with rank 0. Ranks are sometimes also named *levels of surprise*, indicating how much surprising is an event for an involved operator. Therefore, they can be usually specified by a human expert instead of learned from data that may be unavailable.

We have omitted concurrency and mutual-exclusion constraints from this definition in order to avoid excessive complexity and keep our focus on diagnosis. While the causal links in CL, as we defined it, cannot capture all the forms of mutual exclusion, we shall see that implicit mutex constraints play an important role in agents interactions through shared literals. Concurrency and other forms of mutual exclusion, e.g., the use of a resource that has a single instance, could be easily accommodated in our framework.

If we assume that all the actions will be executed in the N mode, a TMAP can be interpreted as a flexible schedule of the plan [14], that guarantees that all the causal links are respected and all the plan actions are smoothly executed. However, the TMAP also contains fundamental information associated with the possible actions failures. In particular, modes different from N are not used for the planning purpose, but for the diagnostic one; such modes allow actions to obtain different effects from the nominal, expected ones.

Example 1. As an example TMAP P, let us consider a case with four agents: $T = \{ag1, ag2, ag3, ag4\}$ (see Fig. 2). The set of actions is $A = \{ac1_1, \ldots, ac4_4\}$, with order relations O and (nominal) causal links CL as shown in the figure, respectively, by the solid and dashed arrows.

Now, we assume that the fifteen actions included in MAP P are instances of just three types of actions: *move*, *load*, and *put*. Intuitively, in the TMAP in

---

[1] For the sake of discussion, we assume that all modes $M(ac)$ of an action ac have the same preconditions *pre*.

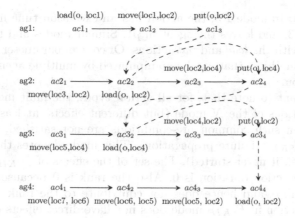

Fig. 2. An example TMAP.

Fig. 2, the four agents have to cooperate for moving an object o from location $loc1$ to location $loc2$ and then $loc4$, and then back to $loc2$ again. For instance, ag_1 moves object o by loading it in $loc1$, and carrying it to location $loc2$. Note the nominal causal link between $ac1_3$ and $ac2_2$, meaning that in a nominal execution of the plan $ag2$ will load block o from $loc2$ after it has been moved there by $ag1$.

The table in Fig. 3 shows the modes $M(ac)$ of each action type, with associated label, pre- and post-conditions, range, and rank.

act	pre	m	post	range	rank
move(ag,p1,p2)	at(ag,p1)	N	at(ag,p2)	[1,3)	0
		F1	at(ag,p2)	[3,10)	1
		F2	at(ag,p2)	[10,25]	2
		F3	∅	[10,25]	3
load(ag,p,o)	at(ag,p), at(o,p), holds(ag,∅)	N	¬at(o,p), holds(ag,o)	[1,2)	0
		F1	¬at(o,p), holds(ag,o)	[2,10)	1
		F2	∅	[10,25]	2
put(ag,p,o)	at(ag,p), holds(ag,o)	N	at(o,p), ¬holds(ag,o)	[1,2)	0
		F1	at(o,p), ¬holds(ag,o)	[2,10)	1
		F2	∅	[10,25]	2

Fig. 3. Example modes.

For instance, in nominal mode (N), a *move(ag, p1, p2)* requires the agent to be in place $p1$, causes the agent to arrive in place $p2$, and has an execution time in the interval $[1, 3)$. The rank is 0, meaning that the N mode is preferred (i.e., the most likely). In mode $F3$, the action has an execution time in the interval $[10, 25]$, a rank 3, and leaves the agent in $p1$. Similar modes and time intervals are associated with the *load* and *put* actions. Of course, parameter o in *load* and *put* represents an object that can be manipulated by multiple agents during the TMAP execution.

It is important to note that, for all action types, the fault modes have the same pre-conditions of the N mode (but different effects, at least in terms of duration). When such common pre-conditions are not satisfied, we assume a special mode F_{SKIP} (failure propagation). This mode denotes that the action was skipped (i.e., it never started). The set of the effects of F_{SKIP} is therefore empty, and the action duration is 0. Also the rank is 0 because it represents a secondary failure, and hence does not contribute to the rank of the overall diagnosis. An action in F_{SKIP} mode does not have direct effects on the world, but may have indirect effects on plan execution since some of its missing effects could be preconditions for subsequent actions.

3   Plan Execution Failure Problem

Timed Observations. We define a timed observation as a pair $\langle e, t \rangle$, where e is the observed event, and t is the time when e occurred. In our TMAP framework, an observable event can be a single ground literal, possibly with a negative polarity. For instance, $at(ag1, p1)$ and $\neg at(ag1, p1)$ are two alternative observable events. Of course, we assume that observations are reliable and consistent (i.e., the same literal does not appear with both polarities at the same time). During the execution of a plan, only a few of these events will be observed (due to *partial observability*).

Plan Execution Failure (PEF) Problem. It is important to note that the agents share the same environment and resources, and cooperate with each other by exchanging services: the effects brought about by an agent may be the preconditions for the actions of another agent. In principle, therefore, the misbehavior of an agent could affect its later activities as well as other agents' activities.

We say that action ac is *ready* when its predecessor ac' s.t. $\langle ac', ac \rangle \in O$ has finished. We assume that, after an action is ready, it will execute as soon as all its pre-conditions are true. In fact, as a consequence of previous failures, the preconditions could be brought about too late, or might even not be provided at all. Let $P = \langle T, A, O, CL, M \rangle$ be a TMAP.

Definition 1. *A mapping $H : A \to M(A) \cup \{F_{SKIP}\}$ is a* hypothesis *about the modes of actions in P that assigns each action $ac \in A$ with a mode $m \in M(ac)$ or special mode F_{SKIP}.*

Since action modes are associated with time intervals and logic pre-/post-conditions, a hypothesis H can be used to estimate a set of possible executions of P, that may differ for the times at which actions start and end; we call these possible executions *temporal execution profiles*.

Definition 2 (Temporal Execution Profile). *Given a TMAP P, and a hypothesis H, a temporal execution profile θ is an ordered sequence of pairs $\langle s_0, t_0 \rangle, \ldots, \langle s_n, t_n \rangle$, such that s_i (i : 0..n) is a state of the whole system consisting of all the atoms holding at time t_i. For each $ac \in A$, the events $T_s(ac)$ (start) and $T_e(ac)$ (end) occur in exactly two states, s_i and s_k, respectively, such that t_i precedes t_k. Moreover:*

1. *each s_i is a set of atoms that are true at time t_i and that represents the state of the whole system*
2. *if ac_j starts at time t_i with mode in $M(ac_j)$, then the preconditions of ac_j for mode $H(ac_j)$ (i.e., the mode assigned by H to ac_j) hold at time t_i, and any other action that starts at time t_i, or is already in progress at that time, is not in conflict with ac_j according to the "no moving targets" rule [5], for which no two actions can simultaneously make use of a value if one of the two is accessing the value to update it;*
3. *if ac_j starts at time t_i with mode F_{SKIP}, then: $t_i = t_k + \tau$ (where t_k is the end time of the predecessor ac_k of ac_j); the preconditions of ac_j do not hold at time t_i; for each $t \in [t_k, t_i]$, if the preconditions of ac_j held at time t, some other action in conflict with ac_j started or was already in progress at time t*
4. *if ac_j ends at time t_i, then the post-conditions of mode $H(ac_j)$ of action ac_j hold at time t_i*
5. *for each action $ac \in A$, the distance between the times when the action starts and terminates belong to m.range where m is $H(ac)$;*
6. *s_0 is the initial given state;*
7. *s_n is the state where the effects of the last performed actions are added.*

Conditions **2** and **4** state that the pre-conditions and effects of an action ac performed with modality $m = H(ac)$ are true, respectively, when the action starts and when the action terminates. Note that condition **2** ensures that two actions that modify the same literal are executed in mutual exclusion; this is a fundamental constraint for actions that affect the value of a shared literal. Condition **3** states that an action is associated with special mode F_{SKIP} only if it has not been allowed to start with true pre-conditions until a timeout τ has expired. Condition **5** imposes that in θ the duration of each action ac respects the intervals of possible durations associated with mode m assumed in H.

Of course, given a TMAP P and a hypothesis H, many temporal execution profiles can be derived: $\mathcal{T}_P(H)$ denotes the set of all possible temporal execution profiles that results from P when only the modalities in H are allowed.

More generally, since each action is associated with a number of modes, we denote with $\mathcal{T}_P$ the space of possible temporal execution profiles for the plan P obtained by considering all possible hypotheses.

Let Obs be a sequence of timed observations over actions in P. Obs can be used as a filter on $\mathcal{T}_P$ by pruning off those profiles that are not consistent with them. More precisely, a temporal execution profile $\theta \in \mathcal{T}_P$ is consistent with Obs iff for each timed observation $\langle e, t \rangle \in Obs$, if we let t_i be the unique time instant in θ such that $t_i \leq t < t_{i+1}$, then $s_i \models e$ (where $\langle s_i, t_i \rangle \in \theta$). In other words, the timed observation $\langle e, t \rangle$ must agree with the state of the world s_i that holds at t according to τ.

It is sufficient that this does not hold for one timed observation in Obs to say that θ is not consistent with Obs. We will denote as $T_P(Obs)$ the subset of the profile space consistent with Obs.

Definition 3 (PEF problem). *A Plan Execution Failure (PEF) problem is a pair $\langle P, Obs \rangle$ where P is a TMAP and Obs a set of timed observations.*

The goal of solving a PEF is to find hypotheses H that are consistent with the observations:

$$T_P(H) \cap T_P(Obs) \neq \emptyset. \tag{1}$$

The previous equation can be expresses in the form of the classic definition of consistency-based diagnosis [15]:

$$P \uplus H \uplus Obs \nvdash \bot.$$

where $\uplus$ represents the intersection of temporal profiles.

It is well known that the number of consistency-based diagnoses can be very large, especially when the observability is low. Therefore, we are not interested in any hypothesis H that satisfies Eq. 1, but only in the hypotheses that also satisfy a preference criterion. More precisely, we look for solutions that minimize the rank (i.e., maximize the probability) associated with the action modes.

Definition 4. *Given a TMAP $P = \langle T, A, O, CL, M \rangle$ and a hypothesis H about actions in P, the rank of H, denoted as $rank(H)$, is*

$$rank(H) = \sum_{ac \in A} H(ac).rank.$$

In fact, since we assign rank 0 to failures that depend on previous failure, and the rank of failures that are independent can be comulated, the rank of a hypothesis is simply the sum of the ranks of the modes assumed in the hypothesis itself. Of course, there exists only one hypothesis H^0 with rank 0 in which all actions are assumed nominal.

Definition 5 (PEF solution). *Let P be a TMAP, and let $\langle P, Obs \rangle$ be a PEF problem, a solution to such a problem is an hypothesis δ such that:*

1. *δ satisfies Eq. (1);*
2. *$rank(\delta)$ is minimal: no other hypothesis H' that satisfies Eq. (1) has $rank(H') < rank(\delta)$*

As usual in a diagnostic setting, we are not interested in just one solution, but in all minimal solutions, in fact, unless other preference criteria are given, all these minimal solutions should be returned as an answer to a PEF problem.

Example 2. Let us consider the plan of Example 1. Although in the original plan, action $put(o, loc2)$ of agent $ag1$ was assumed to make o available for action $load(o, loc2)$ of $ag2$, this may not be the case in a real execution scenario. Assume that the previous action of $ag2$, i.e., $move(loc3, loc2)$, had an $F1$ delay and took 8 time instants. In the meanwhile, the three *move* actions of $ag4$ have taken a total of 6 time instants, so that the object released by $ag1$ at $loc2$ at time 4 is actually loaded by $ag4$. This situation makes actions $ac2_2$, $ac2_4$, $ac3_2$, and $ac3_4$ fail with mode F_{SKIP}, because they don't have the necessary preconditions to be executed. However, a diagnosis that (except for N modes) lists: $ac2_1(F1)$, $ac2_2(F_{SKIP})$, $ac2_4(F_{SKIP})$, $ac3_2(F_{SKIP})$, $ac3_4(F_{SKIP})$ is not a satisfactory *explanation* of what happened. Indeed, the fact that $ac2_1$ had a delay $F1$ does not necessarily imply all the other events and (propagation) failures: think, e.g., that the delay caused by $F1$ was just a duration of 3 time instants for $ac2_1$. In the next section we propose a notion of temporal explanation that yields more information than just the diagnosis.

4 Explaining Failure Propagations

4.1 Temporal Explanations

A solution δ to a PEF problem provides a user with a labeling of (failure) modes to the plan actions that is consistent with the available observations. In particular, a special mode F_{SKIP} in δ is used to denote those actions that have been affected by previously occurred action failures (i.e., it is a *secondary* failure). However, this is not in general sufficient, for the user, to understand what has actually happened. In fact, a secondary failure might be caused by the co-occurrence of two or more primary failures (e.g., when two actions delay independently and their consequences sum up affecting a third action). Such configurations are not easy to discover, and to increase the comprehension of a user, a δ diagnosis needs to be further explained to extract implicit, contingent connections between the primary failure(s) and the secondary ones.

Intuitively, failures can propagate via the shared literals, that is, via the resources produced by an action and consumed by another one. For example, an action may fail because one of the required inputs is not available at the right time, and this may happen because the producer failed in supplying it (including supplying it with too much delay), or because another action has erroneously consumed the resource in its place. Explaining δ, thus, means tracing back the temporal relations among the actions that are related to some resource of interest, and whose occurrence justifies a secondary failure.

Definition 6 (Temporal Explanation of δ w.r.t. R). *Let δ be a PEF solution to $\langle P, Obs \rangle$. A Temporal Explanation (explanation in short) $E(\delta, R)$ of δ w.r.t. a shared literal R is a set of Allen algebra relations among actions in P defined as follows. Let δ_{R+} (resp. δ_{R-}) be the subset of actions in δ that produce (resp. consume) a shared literal R. Moreover, let $\delta_R(F_{SKIP})$ (resp. $\delta_R(\overline{F_{SKIP}})$) be the subset of $\delta_{R+} \cup \delta_{R-}$ containing actions with mode equal to (resp. different from) F_{SKIP}. Then, an explanation $E(\delta, R)$ for δ w.r.t. R is a set such that:*

– *for each ac* $\in \delta_R(\overline{F_{SKIP}})$, $E(\delta, R)$ *specifies two Allen algebra relations* ρ_{prec}
  and ρ_{succ} *w.r.t. its predecessor and its successor in* $\delta_R(\overline{F_{SKIP}})$ *(except for the*
  first and last action). Relation ρ_{prec} *is either* after *or* meets after; *relation*
  ρ_{succ} *is either* before *or* meets;
– *for each ac* $\in \delta_R(F_{SKIP})$, $E(\delta, R)$ *specifies two Allen Algebra during relations*
  ρ_R *(when ac becomes ready) and* ρ_F *(when ac timeouts and fails with* F_{SKIP}*).*
  Relations ρ_R *and* ρ_F *relate ac either with a single action in* $\delta_R(\overline{F_{SKIP}})$, *or*
  with (the interval I in between) two actions ac', ac'' *in set* $\delta_R(\overline{F_{SKIP}})$.

Some comments are in order. First of all, note that, due to the mutual exclusion among actions that produce/consume R, the actions in $\delta_R(\overline{F_{SKIP}})$ respect a total order, specified through the (*meets*) *before/after* relations in $E(\delta, R)$. Such an order partitions the timeline in a set Π_R of intervals of action execution and intervals between two actions.

In addition, an action $ac \in \delta_R(F_{SKIP})$ that was supposed to produce/consume R, but failed because of missing pre-conditions, can actually overlap with actions in $\delta_R(\overline{F_{SKIP}})$. In fact, the action has never started, and what we are interested in knowing is the interval between when ac became ready (i.e., when it became the current action for its agent), and when ac failed with mode F_{SKIP}. Such events, that determine the interval W during which ac is "willing" to produce/consume R are placed in partition Π_R by *during* relations in $E(\delta, R)$. It follows that W is contextualized in $E(\delta, R)$ against all the other intervals regarding the execution of actions that have handled resource R, and hence provides the user with an explanation of why action ac could not produce/consume R during W.

A (full) explanation of a diagnosis δ is simply a set $E(\delta)$ of several sub-explanations $E(\delta, R)$, one for each shared literal R. Note that, given a diagnosis δ, it is in general possible to find several alternative explanations, corresponding to different orders of events compatible with δ. Such alternatives are equally plausible according to our model, and are therefore computed and returned to the human user.

4.2   Explaining Broken Causal Links

As a further refinement, the temporal explanation can be compared against the causal structure of the MAP P induced by the links in CL.

First of all, note that all the links in CL should be satisfied when all the actions of the plan execute in the nominal mode N, i.e., the following property must hold.

Property 1. If we denote as δ^N the hypothesis representing the solution to $\langle P, \emptyset \rangle$, its explanation $E(\delta^N)$ must confirm the links in CL, i.e., if $(ac_i, ac_j) \in CL$ for $ac_i, ac_j \in \delta_R^N$, then $E(\delta^N, R)$ will contain either relation ac_i *before* ac_j or ac_i *meets* ac_j.

This property can be effectively checked with the algorithm for computing explanations that will be given in Sect. 6.

Let us denote by $\gg_{CL}$ the transitive closure of the links in CL. Thus, if two actions ac_i and ac_j are in relation $ac_i \gg_{CL} ac_j$, it means that there exists a chain of causal links and order relations that flow from a_i to a_j. If an action is skipped due to a missing precondition, it will generally be the case that some causal links were broken during the actual execution. By comparing the temporal explanation against the transitive closure on causal links, it is possible to identify these broken links and, to some extent, their causes. Let ac be in $\delta_R(F_{SKIP})$, and let ac_* be in $\delta_R(\overline{F_{SKIP}})$ such that $ac \gg_{CL} ac_*$ according to the definition of the TMAP P. Moreover, let all the causal links in $ac \gg_{CL} ac_*$ be broken, i.e., ac^* is the first action in the chain that is not skipped. As a consequence, all the actions from ac to ac_* did not receive their preconditions as required by CL; however, ac^* did in fact receive its preconditions, although not from the action prescribed by CL. The explanation of δ will clearly state which action provided the pre-condition to ac^*, thus explaining why it succeeded despite the broken causal chain from ac.

The following examples should help clarify the concepts introduced in this section.

Example 3. Let us refer to Example 2. The *producers* of literal $R = at(o, loc2)$ are as follows: $ac1_3$, and $ac3_4$; while the *consumers* of R are: $ac2_2$, and $ac4_4$. According to Example 2, the diagnosis δ (except for N modes) lists: $ac2_1(F1)$, $ac2_2(F_{SKIP})$, $ac2_4(F_{SKIP})$, $ac3_2(F_{SKIP})$, $ac3_4(F_{SKIP})$. The explanation that we have informally sketched in Example 2, should now be formalized as a suitable explanation $E(\delta, R)$. Figure 4 shows $E(\delta, R)$ graphically on a diagram where time increases from left to right. Note that, besides the actions related with R and their Allen algebra relations specified by $E(\delta, R)$ (black), the schema also shows the actions that are affected because they occur along a chain of broken causal links, specifically those falling within the transitive closure $ac2_2 \gg_{CL} ac4_4$, to further increase the information conveyed by the schema to the reader. These actions are in fact assigned with a non-nominal mode by the diagnosis δ and are marked with a $\nearrow$ symbol because they are related to another literal $at(o, loc4)$.

The set of non-F_{SKIP} actions that have to do with R are just $ac1_3$ and $ac4_4$, so that the timeline is partitioned in five regions (dotted vertical bars): before $ac1_3$; during $ac1_3$; between $ac1_3$ and $ac4_4$; during $ac4_4$; after $ac4_4$. The definition of explanation requires us to relate $ac1_3$ and $ac4_4$, and, in the scenario described by Example 2, the relation is $ac1_3$ *before* $ac4_4$, i.e., when $ac1_3$ ends, some time passes before $ac4_4$ becomes ready and consumes $at(o, loc2)$. Two causal links from Fig. 2 are especially in need of an explanation: $(ac1_3, ac2_2)$, because $ac1_3$ has mode N but $ac2_2$ has mode F_{SKIP} (i.e., the action form which the link stems is ok, but the action where the link goes is not); and $(ac3_4, ac4_4)$, because $ac3_4$ has mode F_{SKIP} and $ac4_4$ has mode N (i.e., the "starting" action of the link has failed, but the "ending" one is ok). Both of these facts are explained by the spurious causal link contained in the explanation $E(\delta, R)$, namely $(ac1_3, ac4_4)$, which describes an incorrect actual execution.

Note that R would be available for other consumers between the end of $ac1_3$ and the start of $ac4_4$. However, according to explanation $E(\delta, R)$, $ac2_2$ becomes

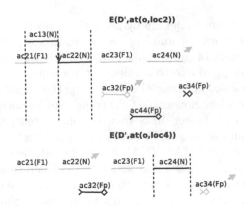

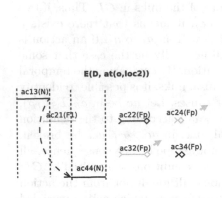

Fig. 4. An explanation of diagnosis $\delta = \{ac2_1(F1), ac2_2(F_P), ac2_4(F_P), ac3_2(F_P), ac3_4(F_P)\}$.

Fig. 5. An explanation of diagnosis $\delta' = \{ac2_1(F1), ac2_3(F1), ac3_2(F_P), ac3_4(F_P), ac4_4(F_P)\}$.

ready and then fails with mode F_{SKIP} (segment starting with > and ending with ◇) only *after* $ac4_4$ ends. By looking at the figure, it is easy to see that such a delay is due to the failure with mode $F1$ of action $ac2_1$.

Let us now consider an explanation for a different diagnosis δ', according to which actions $ac2_1$ and $ac2_3$ fail with mode $F1$ (delay), and actions $ac3_2$, $ac3_4$, and $ac4_4$ fail with mode F_{SKIP} (Fig. 5). The upper part of the figure shows the explanation of δ w.r.t. literal $at(o, loc2)$, while the lower part shows the explanation of δ for literal $at(o, loc4)$. In particular, $ac1_3$ produces literal $at(o, loc2)$, which is consumed by $ac2_2$ (in this case, immediately), as prescribed by the plan. However, we see that although the *put* action $ac2_4$ involving literal $at(o, loc4)$ succeeds, the associated *load* action $ac3_2$ fails with mode F_{SKIP}, which in turn propagates to the failure of actions $ac3_4$ and $ac4_4$ on $at(o, loc2)$ that depend on $ac3_2$ (causal links in the plan, Fig. 2).

By just looking at the upper part of the figure, then, we are left without an explanation of the failure of $ac3_2$. We have to look at the part of the figure showing the explanation for literal $at(o, loc4)$, where we realize that action $ac3_2$ became ready and then failed with mode F_{SKIP} *before* action $ac2_4$ (from which $ac3_2$ depends) was executed. The cause of the delay is clearly a combination of the delays caused by the failures of actions $ac2_1$ and $ac2_3$ with mode $F1$.

5   Translation to SMT

In order to address a PEF problem by exploiting an SMT solver, we have to encode the TMAP and the observations *Obs* in the language accepted by the solver. We recall that an SMT problem is an extension of the well known propositional satisfiability (SAT) problem where formulas can contain relations and functions from various theories including real and integer linear arithmetic. Similar to SAT, when a set of formulas is satisfiable, the solver is able to return a satisfying assignment to the variables. In this work, we have adopted the Z3 solver [3]. Due to space constraints, we will focus just on the most relevant aspects of the encoding process. In order to encode action types (e.g., *move*, *load*), we need to encode the predicates that appear in their pre-conditions and effects, e.g. *at* and *holds*. We define them as *uninterpreted functions* (UF), i.e., functions for which the Z3 solver will try to find an interpretation that satisfies the set of formulas being checked. Note that most of the predicates are in fact *fluents*, i.e., they have time as one of their arguments. For example, *at(ag, p, t)* asserts that agent *ag* is at place *p* at time *t*. For diagnostic purposes, a fundamental predicate is *mode(ac,m)*, that defines the mode *m* of an action *ac*.

Action types, with their behaviors determined by modes, pre-conditions and effects, are expressed as *defined functions* (DF). Unlike UFs, DFs have a body that specifies how to compute the function value given the arguments. A DF receives all the parameters relevant to the action, plus two time points T_s and T_e that represent the action starting and ending times, and returns a Boolean value. For example, the signature of the *move* action is:

$$move(ag : Agent, ac : Action, from : Pos, to : Pos, T_s : Int, T_e : Int) : Bool$$

The body of the DF specifies, for each mode $m \in M(ac)$ and for the special mode F_{SKIP}, the pre-conditions and the effects taken from the TMAP definition.

$$if \ (pre\text{-}cond) \quad mode(ac, N) \Rightarrow [N \ post\text{-}cond]$$
$$\ldots$$
$$mode(ac, F_k) \Rightarrow [F_k \ post\text{-}cond]$$
$$else \qquad mode(ac, F_{SKIP}) \wedge [F_{SKIP} \ post\text{-}cond]$$

The plan itself is encoded as a sequence of assertions that build the instances of action types that make up the plan. Finally, the timed observations *Obs* are easily encoded by asserting the truth of the associated fluent, e.g. the observation $\langle at(ag1, p1), t1 \rangle$ will be encoded by asserting $at(ag1, p1, t1)$.

Constraints between time points are expressed as linear arithmetic relations:

$$T_e(ac) > T_s(ac); \ T_e(ac') < T_s(ac) \leq T_e(ac') + \tau$$

Note that, in the second formula, ac' is the predecessor of ac in the plan of the agent. A fundamental point that needs to be addressed by our translation is the definition of suitable *frame-axioms*, i.e., formulas prescribing that a fluent does not change if none of the actions changing it is taken. For instance, in our example logistic domain, fluent *at(ag, p, t)* is only possibly changed at the end of a *move* action. Moreover, no other agent can change the value of *at(ag, p, t)*. So, for each action *ac* that is not a *move*:

$$at(ag, p, T_e(ac)) = at(ag, p, T_s(ac)) \ \text{ and } \ at(ag, p, T_s(ac)) = at(ag, p, T_e(ac'))$$

where ac' is the predecessor of *ac*. Things are more complicated for fluents such as *at(o, p, t)* (where *o* is an object) that can be changed by multiple agents. According to our assumptions, we impose that such actions must be executed in mutual exclusion. However, in general, for each action ac, we must also assert that:

$$at(o, p, T_s(ac)) = at(o, p, max(\{T_e(ac') : T_e(ac') < T_s(ac)\}))$$

where all actions ac' that can modify the fluent are considered by the $max()$ operator on the right hand side. The encoding of time and persistency relations highlights the benefits of adopting a SMT solver instead of a SAT solver for checking the consistency of hypotheses. In a SAT-encoding of a plan whose timespan is $[0, N]$, it is necessary to create a copy of each variable for each time instant in $[0, N]$. On the contrary, the SMT encoding allows us to focus just on the values of the fluents at the relevant time points, that for a TMAP are the start/end times of actions and the times of observations.

Concurrency constraints are the most difficult ones to encode, especially because we require that an action is actually executed as soon as it is possible to do so. We can't describe in detail such constraints due to lack of space. Suffice it to say that, for each shared literal R in the TMAP P, we need to introduce a predicate $wants(ac, \pm R, T)$, that denotes the fact that an action ac wants to consume $(+)$ or produce $(-)$ literal R at time T. Then, we specify a number of constraints involving the actions $P_R = P_{R+} \cup P_{R-}$ (that produce/consume R) to handle the situations that can arise during plan execution: mutual exclusion, waiting for R to be produced/consumed (possibly competing with other waiting agents), timing out and executing in mode F_{SKIP}.

6   Solving PEF Problems

Given the encoding of a PEF problem in the input language of Z3, we exploit the ability of Z3 to produce an *unsat core* every time it is invoked on an unsatisfiable instance. An unsat core is a set of assertions in the input to Z3 that cannot hold simultaneously and therefore require to withdraw at least one of them in order to get satisfiability. Given the set of unsat cores that is cumulatively produced during the search for the solutions, we can avoid to explore the parts of the search space that do not *hit* (i.e., withdraw at least an assignment from) all of them. This technique is well known in diagnosis, also on approaches based on SMT [7].

Let us assume that we have a function $EncodeTMAPZ3$ that, given a TMAP P, encodes it in the Z3 input language as explained in the previous section. Figure 6 shows the CBFS (Conflict-based Best First Search) diagnostic algorithm for solving a PEF specified by P and *Obs*. The algorithm is strongly based on the high-level schema of Conflict-directed A^* (cd A^*) [18], with some variations explained below.

At each iteration of the top-level *while* loop, algorithm cd A^* would require to generate a full assignment of modes to actions that resolves the conflicts found so far. Instead, we generate a constraint σ on the modes of the actions with function $NextBestPlateauResolvingConflicts()$ (line 5). Such a constraint: (i) contains specific assignments σ_F of faults (excluding F_{SKIP}) to actions in

CBFSDiagnosis(P = ⟨T,A,O,CL,M⟩, Obs)
1. Sys ← EncodeTMAPZ3(P)
2. Pef ← Sys ∪ EncodeObsZ3(Obs)
3. UCores ← ∅; Δ ← ∅; done? ← **false**; best ← ∞
4. **while not** done? **do**
5. σ ← NextBestPlateauResolvingConflicts(UCores)
6.   **if** rank(σ) > best **then**
7.   done? ← **true**
8.   **else**
9.   Pef$_\sigma$ ← Pef ∪ EncodePlateauZ3(σ)
10.   (μ, γ) ← CheckSATZ3(Pef$_\sigma$)
11.   **if** μ ≠ **null then**
12.   best ← rank(σ)
13.   **while** μ ≠ **null do**
14. δ ← project(μ, {$mode(ac,m) \in \mu : ac \in A$})
15. Δ ← $\Delta \cup \{\delta\}$
16.   Pef$_\sigma$ ← Pef$_\sigma$ ∪ EncodeAssignmentZ3($\neg\delta$)
17.   (μ, γ) ← CheckSATZ3(Pef$_\sigma$)
18.   **end while**
19.   **else**
20.   UCores ← UCores ∪ γ
21.   **end if**
22.   **end if**
23. **end while**
24. **return** Δ

Fig. 6. The CBFS diagnostic algorithm.

order to hit all the unsat cores $\gamma \in UCores$; (ii) constrains the remaining actions to have either mode N or F_{SKIP}; (iii) has minimum rank among assignments that hit $UCores$. Therefore, σ looks as follows:

$$\sigma = ac_1^F(\varphi_1) \wedge \ldots \wedge ac_m^F(\varphi_m) \wedge$$
$$(ac_1^{R0}(N) \vee ac_1^{R0}(F_{SKIP})) \wedge \ldots \wedge (ac_n^{R0}(N) \vee ac_n^{R0}(F_{SKIP}))$$

where actions $ac_i^F \in \sigma_F$ are assigned a specific faulty mode φ_i (excluding F_{SKIP}), while actions ac_i^{R0} (where the superscript $R0$ denotes the fact that such actions contribute a rank 0 to the assignment) can take mode N or F_{SKIP}. This explains the term *plateau* in the name of the function that computes constraints σ: a single constraint may indeed generate several diagnoses of equal rank (i.e., cost) by assigning combinations of modes N or F_{SKIP} to the ac_i^{R0} actions (see below).

When all the minimum rank solutions Δ to a given TMAP have already been found, a constraint σ with a higher rank than the *best* one is generated (line 6), and the algorithm returns set Δ. Otherwise, the constraint σ is added to the Z3 encoding *Pef* of the PEF problem (TMAP and observations), and the result

Explain(Pef, δ)
1. Pef$_\delta$ ← Pef ∪ EncodeAssignmentZ3(δ)
2. (μ, γ) ← CheckSATZ3(Pef$_\delta$)
3. **while** $\mu \neq$ **null do**
4. e_{raw} ← project(μ, $\{T_s(ac) \in \mu : ac \in A\}$)
5. e_{All} ← EncodeAllenAlgebra(e_{raw})
6. E ← $E \cup e_{All}$
7.   Pef$_\delta$ ← Pef$_\delta$ ∪ ¬e_{All}
8. (μ, γ) ← CheckSATZ3(Pef$_\delta$)
9. **end while**
10. **return** E

Fig. 7. The Explain algorithm.

Pef$_\sigma$ is then checked by Z3 for satisfiability. If Pef$_\sigma$ is unsatisfiable, Z3 returns an unsat core γ, that is added to the set *UCores*.

Otherwise, a satisfying model μ is returned by Z3. The *best* rank of solutions is updated with the rank of σ_F. Then, the algorithm enters an inner while loop where: the full assignment δ to the action modes prescribed by μ is added to the set Δ of preferred diagnoses; and then Pef$_\sigma$ is checked again for satisfiability excluding δ (to avoid finding it again).

The explanations of a diagnosis δ are computed with the *Explain* algorithm shown in Fig. 7. Diagnosis δ is added to the encoding *Pef* of the PEF problem solved by δ, and the result *Pef$_\delta$* is checked for satisfiability with Z3. Of course, since δ is a diagnosis, the while loop is entered at least once. The times of start and end of each action are extracted from model μ, and then they are abstracted into set e_{All} of the corresponding Allen algebra relations introduced in Definition 6. For example, if a *put(ag, p, o)* ends at time t, and a *load(ag', p, o)* starts at time $t+1$, then a relation *meets* is established between *put* and *load*. After adding e_{All} to the set E of explanations of δ and negating it in Pef$_\delta$, Z3 is called again to look for other explanations of δ.

7 Implementation and Results on Test Cases

We have implemented the SMT-based approach to diagnosis described above as a Java program exploiting the Z3 solver. The tests have been run on a machine running Ubuntu 18.04.1 LTS, equipped with an i7 7700HQ CPU at 2.80 GHz, and 8 GB RAM. We have considered a *Logistic* domain which reflects the domain used in the examples. In such a domain, as mentioned above, agents can *move*, *load*, and *put* objects, giving rise to several kinds of inter-agent interactions. We have experimented our approach by running a number of software simulated tests under different *configurations*, defined by varying the following dimensions: *#ag* (2 and 4 agents); *#ac* (8, 10, 20 actions per agent); and *#rnk* (injected failures of ranks 2, 4). In order to study the effect of interactions among agents,

Table 1. avg time (sec), sols, time/sol, and explanations of experiments.

	CBFS			
	Time	#sol	Time/sol	#expl
ag 2				
ac 8 (R2)	0.48	2.0	0.24	2.0
ag 4				
ac 10 (R2)	1.32	2.5	0.53	3.0
ac 20 (R2)	6.83	4.0	1.71	6.1
ac 20 (R4)	25.53	15.6	1.64	23.2

we have introduced inter-agent links in the plans used in the configurations as follows: 2 ag × 8 act with 2 links; 4 ag × 10 act with 3 links; and 4 ag × 20 act with 7 links.

The observability rate (i.e., ratio between the number of actions with observable effects and the total number of actions) was 30%. We have chosen this level of observability because it has proved to be high enough for our algorithm to (almost) always include the diagnosis with the injected failures in the list of preferred diagnoses, and low enough to challenge our algorithm with a certain ambiguity in discriminating between the "real" diagnosis and alternative ones.

In Table 1, we show results obtained with 4 different configurations of increasing complexity. The average total time for solving the PEF problems goes from 0.48 s (2 agents × 8 actions, rank 2), up to 25.53 s (4 agents × 20 actions, rank 4). It should be noted that the total time includes the computation of all the preferred diagnoses, as well as their temporal explanations. If we look at the average time taken for computing each preferred diagnosis (including its explanations), the increase is more limited, going from 0.24 s to 1.71 s. Indeed, as the test cases become more challenging (more agents, more actions, higher rank of failures), the average number of preferred diagnoses increases (from 2.0 to 15.6), as well as the average number of associated explanations (from 2.0 to 23.2). Note that the time/sol of the 3rd and the 4th configurations is almost the same, despite the fact that the former has test cases with rank 2 and the latter of rank 4. This seems to indicate that the time/sol is not affected significantly by the rank of test cases.

8 Conclusions

The diagnosis of Temporal Multiagent Plans (TMAPs) has been addressed by a number of approaches in literature that focused either on diagnosing delays [12,13,16], or on diagnosing violation of logic conditions [9–11]. In this paper, we have presented a novel approach that deals with both aspects. As a consequence, the propagation of failures from one action to another (and one agent to another one) is particularly complex, because it can be due to delays and/or missing logic effects. Therefore, in our framework we first single out diagnoses (possibly containing secondary failures) by means of a conflict-based search. We then explain these secondary failures by inferring the temporal profile of the production/consumption of shared resources whose misuse caused the very same failures. These temporal profiles allow a user to gain a better understanding about the causes of a secondary failure by relating it to the (primary) failure of another action that has caused an unexpected effect on some shared resource.

Some recent works in the literature address the explanation of the behavior of agents whose internals are based on "black-box" components, mostly realized through Machine Learning and/or Data Mining techniques [1,8]. Contrary to such works, we can exploit a quite accurate model of our system (i.e., the TMAP that is being executed); in this sense, our approach is more closely related to the approaches to explainable planning discussed in [1]. As witnessed by those papers, as well as by the present one, the availability of an intelligible model does not imply that conveying a clear and intuitive explanation of an intelligent task to the human user is trivial.

We are considering several future extensions of the present work. Currently, action failures in plan execution are considered as independent of one another, except when the actions interact through shared literals. Following [13], we may try to extend the present work by considering that action failures can be related also when they involve some common features of the agent or the environment (e.g., a motor or a traffic jam for a *move*). Since plan diagnosis is the precondition for plan repair, another future line of work will explore how to exploit the (on-line) computation of diagnoses to inform a re-planning process that tries to achieve (most of) the original goals.

References

1. Adadi, A., Berrada, M.: Peeking inside the black-box: a survey on explainable artificial intelligence (XAI). IEEE Access **6**, 52138–52160 (2018)
2. Anjomshoae, S., Najjar, A., Calvaresi, D., Främling, K.: Explainable agents and robots: results from a systematic literature review. In: Proceedings of the 18th International Conference on Autonomous Agents and Multiagent Systems, AAMAS 2019, Montreal, QC, Canada, 13–17 May 2019, pp. 1078–1088 (2019)
3. de Moura, L., Bjørner, N.: Z3: an efficient SMT solver. In: Ramakrishnan, C.R., Rehof, J. (eds.) TACAS 2008. LNCS, vol. 4963, pp. 337–340. Springer, Heidelberg (2008). https://doi.org/10.1007/978-3-540-78800-3_24
4. Elimelech, O., Stern, R., Kalech, M., Bar-Zeev, Y.: Diagnosing resource usage failures in multi-agent systems. Expert Syst. Appl. **77**, 44–56 (2017)
5. Fox, M., Long, D.: PDDL2.1: an extension to PDDL for expressing temporal planning domains. J. Artif. Intell. Res. (JAIR) **20**, 61–124 (2003)
6. Goldszmidt, M., Pearl, J.: Rank-based systems: a simple approach to belief revision, belief update, and reasoning about evidence and actions. In: Proceeding of the KR 1992, pp. 661–672 (1992)
7. Grastien, A.: Diagnosis of hybrid systems with SMT: opportunities and challenges. In: Proceedings of the 21st European Conference on AI, pp. 405–410. IOS Press (2014)

8. Guidotti, R., Monreale, A., Ruggieri, S., Turini, F., Giannotti, F., Pedreschi, D.: A survey of methods for explaining black box models. ACM Comput. Surv. (CSUR) **51**(5), 93 (2018)

9. de Jonge, F., Roos, N., Witteveen, C.: Primary and secondary diagnosis of multi-agent plan execution. Auton. Agents Multi-Agent Syst. **18**, 267–294 (2009)

10. Kalech, M., Kaminka, G.A.: On the design of coordination diagnosis algorithms for teams of situated agents. Artif. Intell. **171**(8–9), 491–513 (2007)

11. Micalizio, R., Torasso, P.: Cooperative monitoring to diagnose multiagent plans. J. Artif. Intell. Res. **51**, 1–70 (2014)

12. Micalizio, R., Torta, G.: Diagnosing delays in multi-agent plans execution. In: Proceedings of the 20th ECAI, pp. 594–599. IOS Press (2012)

13. Micalizio, R., Torta, G.: Explaining interdependent action delays in multiagent plans execution. Auton. Agents Multi-Agent Syst. **30**(4), 601–639 (2016)

14. Policella, N., Smith, S.F., Cesta, A., Oddi, A.: Generating robust schedules through temporal flexibility. In: ICAPS, vol. 4, pp. 209–218 (2004)

15. Reiter, R.: A theory of diagnosis from first principles. Artif. Intell. **32**(1), 57–95 (1987)

16. Roos, N., Witteveen, C.: Diagnosis of simple temporal networks. In: Proceedings of ECAI 2008, pp. 593–597 (2008)

17. Torta, G., Micalizio, R., Sormano, S.: Explaining failures propagations in the execution of multi-agent temporal plans. In: Proceedings of the 18th International Conference on Autonomous Agents and Multiagent Systems, AAMAS 2019, Montreal, QC, Canada, 13–17 May 2019, pp. 2232–2234 (2019)

18. Williams, B.C., Ragno, R.J.: Conflict-directed A* and its role in model-based embedded systems. Discrete Appl. Math. **155**(12), 1562–1595 (2007)

Explainable Argumentation for Wellness Consultation

Isabel Sassoon(✉), Nadin Kökciyan, Elizabeth Sklar, and Simon Parsons

Department of Informatics, King's College London, London WC2R 2LS, UK
{isabel.k.sassoon,nadin.kokciyan,elizabeth.sklar,simon.parsons}@kcl.ac.uk

Abstract. There has been a recent resurgence in the area of explainable artificial intelligence as researchers and practitioners seek to provide more transparency to their algorithms. Much of this research is focused on explicitly explaining decisions or actions to a human observer, and it should not be controversial to say that looking at how humans explain to each other can serve as a useful starting point for explanation in artificial intelligence. However, it is fair to say that most work in explainable artificial intelligence uses only the researchers' intuition of what constitutes a 'good' explanation. There exist vast and valuable bodies of research in philosophy, psychology, and cognitive science of how people define, generate, select, evaluate, and present explanations, which argues that people employ certain cognitive biases and social expectations to the explanation process. This paper argues that the field of explainable artificial intelligence can build on this existing research, and reviews relevant papers from philosophy, cognitive psychology/science, and social psychology, which study these topics. It draws out some important findings, and discusses ways that these can be infused with work on explainable artificial intelligence.

Keywords: Explanation · Explainability · Interpretability · Explainable AI · Transparency

1 Introduction

1.1 Scope

In the work presented here, we explore the application of *computational argumentation* and *argumentation-based dialogue* to the domain of *clinical consultation*, particularly focusing on patient self-management of chronic health conditions. Such conditions are characterised by regular and sometimes frequent monitoring of various biometric signs, along with a prescribed regimen of diet, exercise and medication. A variety of different types of exchanges may occur between a patient and their health care provider(s). Here, we propose an agent-based system designed to support consultations that a patient may engage in as a supplement to typical periodic appointments with general practitioners and other professional health care providers that are required of patients with chronic

© Springer Nature Switzerland AG 2019
D. Calvaresi et al. (Eds.): EXTRAAMAS 2019, LNAI 11763, pp. 186–202, 2019.
https://doi.org/10.1007/978-3-030-30391-4_11

health conditions. We consider the types of interactions that a patient might have with an agent-based system that supports the patient's needs for such consultations, and we have identified three key functionalities that such a system could provide: (a) information, (b) recommendation, and (c) explanation.

1.2 Wellness Consultations

A patient with chronic and/or multiple medical conditions will rely on periodical and regular interactions with their General Practitioner (GP), however there may be some questions or advice that is required that does not need to rely on a face to face meeting with their GP. These interactions would be complementary to regular contact with GPs but would offer the opportunities for some advice or recommendations to be obtained sooner. It is this type of asynchronous interaction through a dialogue that we define as a *wellness consultation* throughout this paper.

The first key function involves the patient querying the system and requesting *information* about their condition—for example, details about symptoms, treatment options and drug interactions—the kinds of information that appear in brochures provided by GPs, other health care professionals (HCPs) or pharmacists. This is characterised by a two-step interaction in which the patient initiates a query and the system retrieves the answer from a knowledge store and presents it to the patient.

The second function involves the patient asking the system to *recommend* a course of action, such as engaging in exercise activities tailored to the patient's condition, considering an over-the-counter medication such as *ibuprofen* to alleviate pain, or seeking further advice from an HCP. The distinction from the first function is an expectation that the patient will follow up their interaction with the system with some type of action; and so the system may later ask the patient if they have performed that action in order to help track their condition. While the initial query is initiated by the patient, the system could initiate the follow-up query to check later, e.g., if the patient has taken the recommended painkiller and is still experiencing pain or if the patient has taken the recommended exercise.

The third function involves the patient asking for clarification or *explanation* about a response given by the system in either of the first two cases. Where the first two functions are characterised by two-step interactions (one party initiates and the other responds), the third function involves an iterative process in which several queries and responses occur until the patient understands the information and/or recommendation offered by the system.

The first function is not an uncommon feature in today's IT[1]-rich society, with many people around the world being connected on mobile (and desktop) devices to internet sites that can offer health information. The second function is less common as a commercial product, but is certainly the focus of many IT research projects (including the one that led to this paper and those mentioned

---

[1] IT: Information Technology.

		software agent	
		patient	GP
human user	patient	self-care	consultation
	GP	training	second opinion

Fig. 1. The different roles that a *human user* and *software agent* might play in an agent-based system designed to support self-management of chronic conditions.

in Sect. 5). The third function is even less common, both in commercial and research forms. The notion of *Explainable AI* has become a trend relatively recently (although many of the ideas and motivation behind Explainable AI are not so new, as pointed out in Sect. 5). The work presented here focuses on the third functionality—*explanation*—and proposes the use of *computational argumentation* and *argumentation-based dialogue* (see Sect. 2) as the means to implement the *explanation* functionality.

Consider a wellness consultation involving a conversation between a GP and a patient, where the patient is either being offered advice or is asking for some. Imagine an interactive agent-based system that could facilitate this conversation. Hypothetically, this "agent" could take either role: i.e., that of the patient or the GP, as illustrated by Fig. 1.

When the human user is a patient and the agent acts as a GP, then the interaction is an example of a *consultation* situation, as we have described above. When the human user is a patient and the agent also acts as a patient, then the interaction is akin to a conversation between peers in which the agent could offer reminders and encouragement regarding *self-care*. When the human user is a GP and the agent acts as a patient, then the interaction can be seen as a form of *training*, where the GP user could practice conversing with patients in order to confirm or practice the diagnostic process. When the human user is a GP and the agent also acts as a GP, then the interaction is similar to that of seeking a *second opinion*, or a case review. Note that these scenarios make assumptions about differing levels of expertise corresponding to the two possible user roles, assuming, with respect to medical knowledge, a relatively naïve patient and a more educated GP.

1.3   Contributions

In this paper, we explore the dialogue types relevant to wellness consultations as motivated by the CONSULT project[2], and propose dialogue templates for a set of types of interactions between the CONSULT system and the patient, through the use of a chatbot. In order to do this, we first explore the wider context of interactions between a human and an agent through dialogue, and use the framework introduced in [30] to map the possible interactions within Clinical Decision Support in general, and the CONSULT project in particular. Hence, in this initial approach we focus on the types of interactions between an agent

---

[2] https://consult.kcl.ac.uk.

acting as the GP, and the patient as the human in the wellness consultation dialogue (This interaction is in the bottom left quadrant of the matrix in Fig. 1).

The contributions of this paper are as follows: (1) we show how wellness consultations can be supported by existing types of human-agent argumentation-based dialogues; (2) we articulate how these argumentation-based dialogues can provide explanations, in particular by using a novel combination of argumentation schemes and explanation templates; and (3) we show the applicability of our approach on a clinical example from the CONSULT project. The paper is structured as follows: Sect. 2 provides background on argumentation, argumentation based dialogue and explainable AI. In Sect. 3, we outline our proposed approach. We then use an example from the CONSULT project in Sect. 4 to illustrate our approach. Section 5 briefly discusses related work and Sect. 6 summarizes the work and outlines our plans for future research.

2   Background

2.1   Computational Argumentation

Computational Argumentation [27] is a well-founded method which allows reasoning with incomplete and at times conflicting information or knowledge. An argument is structured so that it has a conclusion or a claim, and the support for the claim. When argumentation is employed in the context of decision support, its structure supports a human-like reasoning process where arguments' conclusions, their support, and the relationships between them, can be modelled. Argumentation has been extensively explored within the multi-agent community, and there are examples of its application to clinical decision support, which is the domain this work is focusing on. In Sect. 5 we discuss some of these applications.

An argument, $Arg = \langle S, c \rangle$, consists of a set of premises, S, defined in some language, $\mathcal{L}$, which support the conclusion, c. An *Argumentation Framework (AF)* [7] represents a set of Arguments $\mathcal{A}$, and the relationships between the members of the set. Formally an AF is a pair $\langle \mathcal{A}, \mathcal{R} \rangle$, where $\mathcal{A}$ is a set of arguments and $\mathcal{R}$ is binary relation representing *attack* relationships between arguments. For example if $Arg_1 = \langle p_1, c_1 \rangle$ and $Arg_2 = \langle p_2, \neg c_1 \rangle$ then an attack relation exists between Arg_1 and Arg_2 since these arguments have conflicting conclusions (i.e. rebuttal attack). Given a set of conflicting arguments, there are well-founded methods [7] for computing *extensions*, consistent sets of arguments which represent coherent opinions.

2.2   Argumentation-Based Dialogue

In their seminal work, Walton and Krabbe [32] described six primary types of dialogue: *information seeking, inquiry, persuasion, negotiation, deliberation* and *eristic*. The distinguishing factors amongst these different types of dialogue are based on a participant's knowledge, their individual goals and goals that they share with others. A logic-based formalism for modeling such dialogues as a formal *game* between agents was introduced in [16] and extended in [30]. This formalism supports the combination of multiple dialogues. [22] examines a subset of these dialogue types and shows how their properties depend on the behaviour of the agents engaging in the dialogue. The different types of dialogues include:

- *information seeking* [32], where one participant asks a question that she does not know the answer to and believes the other participant can answer;
- *inquiry* [15], where both participants seek an answer to a question that neither knows the answer to;
- *persuasion* [24], where one participant tries to alter the beliefs of another participant;
- *negotiation* [26], where participants bargain over the allocation of a scarce resource;
- *deliberation* [17], where participants decide together on taking a particular action;
- *eristic* [32], where participants quarrel verbally;
- *command* [9], where one participant tells another what to do;
- *chance discovery* [13], where a new idea arises out of the discussion between participants; and
- *verification* [5], where one participant asks a question that she already knows the answer to and she believes the other participant also knows the answer, so her aim is to verify her belief.

In the context of wellness consultation interactions it can be seen that not all dialogue forms will be relevant. The types of dialogues of most relevance are those related to deliberation and persuasion. An application of some of these dialogue types in the context of human robot interaction is outlined in [30].

2.3   Argumentation Schemes

An *argumentation scheme or argument scheme* (AS) is a semi-formal reasoning template that matches common reasoning patterns in real life. An *Argumentation Scheme* is a model for instantiating arguments within a specific context and is used to provide a formal basis for instantiating arguments and defining their internal structure. An AS consists of a set of support premises (S), which support the conclusion premise, c, necessary for this derivation [20,21,23,33].

Formally: An argument scheme $AS = \langle S, c, \mathcal{V} \rangle$ consists of the set of premises, S, which support a conclusion, c, and are instantiated with the set of variables, $\mathcal{V} = S.V \cup c.V$.

One of the key features of argumentation schemes is the list of associated critical questions (CQs). The claim or conclusion that the scheme supports is presumptive and the claim is withdrawn unless the CQs posed have been answered successfully. The instantiation of the appropriate argumentation scheme and its associated CQs is a way of generating a set of arguments that can then be reasoned with as an argumentation framework. This mechanism ensures that only arguments that have not been defeated by the CQs will be generated, furthermore CQs can also point to additional arguments to consider.

3   Our Approach

Our approach to supporting explanation functionality as part of a wellness consultation between a human and an agent involves three steps: (1) defining an *argumentation scheme* specific to the provision of health related treatments or actions; (2) identifying existing, and possibly defining new, *argumentation-based dialogues* that are appropriate for this domain; and (3) showing how these schemes in conjunction with the argumentation-based dialogues can be used by the agent to provide *explanation* to patients (human users). In this section, we describe each of these elements to our approach. Then following, Sect. 4, we provide concrete examples of how our approach is (will be) implemented.

Our approach relies on a method to generate recommendations about a suitable treatment or action required to attain a clinical goal. For this purpose we propose the use of a clinically specialized argumentation scheme in order to instantiate and reason with all the possible relevant options. Reasoning with set of generated arguments can be achieved using a variety of approaches such as [7], which will result in an extension containing arguments in support of a treatment or an action. These will then form the basis of the elements of the dialogues and will leverage a set of explanation templates to ensure they can be communicated clearly to the user. The dialogue protocols we propose dictate the order and options the dialogue can progress in.

3.1   Argumentation Scheme for Proposed Treatment (ASPT)

The domain of relevance to this approach involves recommending a course of action in the clinical context, we therefore need to employ an argumentation scheme that is specialized to this domain. We propose to use the *Argumentation Scheme for proposed treatment (ASPT)* [11], which is a specialization of *argumentation scheme from practical reasoning ASPR* [1]. There are undoubtedly additional argumentation schemes that can be used to support dialogues in the clinical context, in the future we will explore the use of additional schemes as part of dialogues related to wellness consultations.

Within our proposed approach the wellness consultation between a patient and the agent to deliberate about possible actions or treatments is underpinned by ASPT. ASPT is in Fig. 2. The arguments instantiated by ASPT are in support of different treatments or actions.

ASPT
$p1$ - Given the patient Facts F
$p2$ - In order to realise the goal G
$p3$ - Treatment T promotes the goal G
therefore : Treatment T should be considered

Fig. 2. Argumentation scheme for a proposed treatment [11].

ASPT is subject to a set of critical questions (CQs), these can be the source of additional or counter-arguments to the arguments instantiated by this scheme. There is no general agreed method for structuring AS and their CQs as such when outlining ASPT and its CQs we made use of the applicable premises and CQs from ASPR as a starting point and specialized them to this clinical setting. We also made a distinction in separating the premises of the scheme and the critical questions by the nature of the information they require. The premises of the scheme rely on facts and information unlikely to change such as clinical guidelines or a specific patient's demographics, whilst critical questions look more at the patient's specific clinical history. The list of critical questions outlined herein is a subset of the CQs proposed for this Argument Scheme. We are using this subset of three critical questions to illustrate our approach, as the formalization of ASPT is not the focus of this paper. The subset of critical questions for the argument scheme ASPT are in Fig. 3.

Critical Questions for ASPT
[CQ1] Has treatment T been unsuccessfully used on the patient in the past?
[CQ2] Has treatment T caused side effects on the patient?
[CQ3] Given patient facts F, are there any counter-indications to treatment T?

Fig. 3. Critical questions for ASPT [11].

Given a goal that needs to be realised, instantiating ASPT and its critical questions will result in an Argumentation Framework. Reasoning with this framework will generate extensions that contain treatment suggestions. Such an approach is outlined in [11] and [2]. When exploring options to explain any recommendations made, then all the relevant arguments and their critical questions will be possible sources of explanations.

The aim of the dialogue as part of a wellness consultation is for the agent to propose an action or treatment to the patient, and to allow the patient to query the rationale underlying the recommendation made, if they wish to do so. In the next section we model the possible situations that will provide a framework for the possible dialogues required. These rely on the beliefs that the human will have in respect to the action or treatment proposed by the agent.

3.2   Dialogue

In the context of a wellness consultation, we assume that when a dialogue takes place it will be regarding a specific action, for example one supported by an argument instantiated by ASPT. We assume that the goal that underlies the recommendation is mutually agreed upon by the human and the agent. In other words the dialogue is not about the goal, but about the actions to take to achieve the goal. This is a strong assumption because, in general, there will be many interactions in which it is necessary to discuss the goal—the agent may need to elicit the goal from the human or persuade them to adopt a goal. However, we think it is a reasonable assumption in our current work since we are only focused on providing recommendations when the human specifically asks for one—it seems reasonable that a user who has asked for a recommendation for a course of action will already have committed to taking that course of action—and we defer considering more general scenarios to future work.

The agent has a known set of beliefs $Ag.\Sigma$, and the human may also have a set of beliefs. $Ag.\Gamma(H)$ represents the agent's beliefs about the human's belief. So far this uses a notation similar to [30]. b represents a belief about an action, this can be an action recommended by instantiating ASPT. For example in case of treating a condition with a specific drug then the action is $b = \mathbf{offer}(arb, hbp)$ (where arb is a treatment for *high blood pressure (hbp)*). $\neg b$ is a disbelief in that action, and $?b$ is a situation where there is no information about b. The possible dialogues that can occur in the context of a wellness consultation when the agent is the GP and the human is the patient are in Fig. 4.

	$b \in Ag.\Gamma(H)$	$\neg b \in Ag.\Gamma(H)$	$?b \in Ag.\Gamma(H)$
$b \in Ag\Sigma$	case 1 agreement - (no dialogue)	case 2 disagreement - (persuasion)	case 3 agreement + explanation - (information seeking)
$\neg b \in Ag\Sigma$	case 4 disagreement - (persuasion $\rightarrow$ deliberation)	case 5 agreement - (deliberation)	case 6 (deliberation)
$?b \in Ag\Sigma$	case 7 (deliberation)	case 8 (deliberation)	case 9 (deliberation)

Fig. 4. The space of possible dialogues between agent and user. The agent, Ag, assumes the role of GP and the user, H, is the patient.

The following situations are relevant in the context of an agent acting as the GP and the human being a patient:

- *Agreement*: Both parties' beliefs do not conflict. There is no need for a dialogue on the specific action in question. However if the agreement is against the action (i.e. both agree $\neg b$), and there is a need to consider an alternative action then this may lead to a deliberation dialogue (case 5).
- *Disagreement*: Beliefs are in conflict (such as in case 2 and 4), the type of dialogue to initiate depends on the user's expertise. If the user is an expert (another GP) or a layperson in medicine (a more plausible scenario) then the dialogue burden of proof and dynamics will vary. In this initial approach we assume that the patient's expertise is less trusted compared to the GP, therefore in this case the agent (as the GP) will initiate a persuasion dialogue. However there may be different levels of disagreement such as the patient not just disagreeing with the proposed action, but proposing a different action instead.

– *Deliberation (referral)*: If the agent does not believe in the action being discussed or has no information on it, then this requires an alternative action to
be agreed upon. Hence a deliberation dialogue, which may require a nested
persuasion dialogue [14].

The dialogue flow is made up of three basic building blocks, an overall *deliberation* dialogue, and depending on the locutions of the human, this can then nest
either a *persuasion* dialogue or a *explanation (or follow up) dialogue* dialogue.
The latter two are similar but their suitability to a given situation depends on
whether the locution of the human was one of agreement/acceptance (in which
case this is a *explanation (follow-up) dialogue*). Should the human locution be
confrontational then this would be more suitably addressed by a *persuasion dialogue*. This flow is illustrated in Fig. 5.

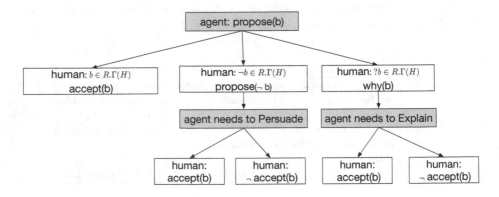

Fig. 5. The possible dialogue options matching the top row of Fig. 4

3.3   Explanation Templates

An important element of our approach is the *explanation* functionality as part of
the *Argumentation-based Dialogue*. We propose the use of *explanation templates*
which will be defined for the argument scheme as well as for its critical questions.
These templates are specific to the reasoning and specialization of the scheme,
and include placeholders for the actual instantiated variables specific to a given
application of the scheme. A sample set of templates are given in Fig. 6.

These templates work as follows. Given a goal G, patient facts F and a
knowledge base of treatments, when a dialogue results in the instantiation of
ASPT, the relevant critical questions are also instantiated, resulting in an argumentation framework that can be solved to establish a set of extensions. Each
element in an extension is mapped to their source (the AS or CQ it was derived
from) and their instantiated values. The templates in Fig. 6 can then be used to
create explanations—for example if a user queries why a particular treatment T
is recommended, the explanation template for ASPT can be instantiated with
the relevant values of G and F (see below).

AS or CQ	Explanation template
ASPT	Treatment T should be considered as it promotes goal G, given patient facts F
CQ1	Treatment T should not be considered as it was not effective for this patient in the past
CQ2	Treatment T should not be considered as it caused side effects for this patient in the past
CQ3	Treatment T should not be considered as patient fact $f_i \in F$ is a counter-indication to its use

Fig. 6. Mapping of argument schemes and critical questions to explanation templates.

4   Example from the Medical Domain

In order to illustrate the approach proposed in this paper we make use of an example interaction between the agent and the patient that arises in the context of the CONSULT project. The aim of the CONSULT project is to develop and test the feasibility of a collaborative mobile decision support system to help patients who suffer from chronic diseases and multiple co-morbidities to manage their treatment. The prototype system integrates data from wellness sensors, electronic health records and relevant guidelines to support data-backed argumentation based decision support. This is accessible to the patient via a mobile app that includes a dashboard and a chatbot. In this example we assume that the agent is acting the role of the GP, and the human is the patient (as outlined in Fig. 1).

Patient uses chatbot	Patient asks a question about symptom s_1, what action to take?

	Agent maps s_1 to goal g
	Agent instantiates argumentation engine and aspt(g,...,)
	Extension includes argument offer(b)
Agent uses chatbot	Agent proposes action b using chatbot

Fig. 7. The steps to start the dialogue when a patient asks for advice

4.1   Patient Asks for Advice

This interaction will be one initiated by the patient when they ask for a recommendation, and we will assume we have a fictitious patient called Bob. In this scenario the patient (Bob) asks a question or seeks advice regarding a specific symptom s_1 via a chatbot dialogue. The initial processes generated by this are in Fig. 7. Figure 7 illustrates the steps taken before the dialogue commences. These are the mapping of the symptom $s1$ to a goal g, the instantiation of the Argumentation Scheme for Possible Treatment (ASPT) using the relevant variables (one of which is g). The mappings between symptoms and goals are assumed to be in the agent's knowledge base. Reasoning with all the arguments results in an extension including an action (for example b) will form the basis for the dialogue. A view of a dialogue options that are possible after a recommendation is made by the agent is depicted in Fig. 8. Note that we assume the argumentation engine constructs an argument based on its knowledge base and the information it received from external sources, such as the patient's EHR, sensor data and relevant clinical guidelines.

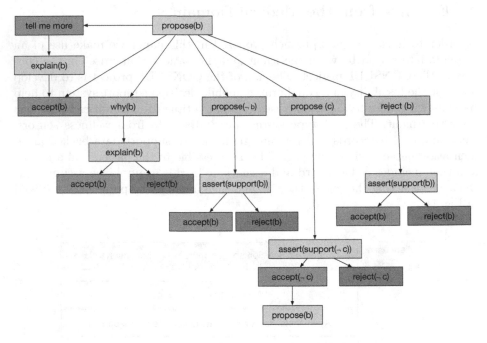

Fig. 8. A possible "patient asks for advice" dialogue tree

In this dialogue when Bob asks for a recommendation, Bob uses the agent to self manage pain by initiating a wellness consultation. The protocol for this dialogue, split into its possible branches, for this dialogue is illustrated in Figs. 9, 10 and 11. This illustration of dialogue protocol is the same format used in [30]. These Figures include the different possible ways in which the dialogue can evolve and the elements of the arguments, critical questions and explanations used within these.

The instantiation of ASPT and its related CQs in this case results a recommendation for *ibuprofen* as a painkiller. The agent proposes the recommended action to Bob (Fig. 9), and Bob can respond in a few ways. Bob can: simply accept the recommendation; ask for more information (Fig. 10); reject the recommendation (Fig. 10); or propose an alternative action (Fig. 11).

Should Bob ask for more explanations (by asking why) then this would trigger an explanation dialogue that would initially outline the reasoning within the argument in support of the use of *ibuprofen* using the template in Fig. 6. An example dialogue within the flow of Fig. 10 where Bob asks for more information (The elements in parenthesis are not part of the actual dialogue):

- *agent*: It is recommended that you take Ibuprofen
- *bob*: Why should I take Ibuprofen?
- *agent*: Ibuprofen (Treatment T) should be considered as it promotes back pain relief (goal G) given your clinical history (Bob patient facts F).

Here the agent is instantiating the explanation template from Fig. 6 that matches ASPT.

Bob would then be able to accept the recommendation, or probe further. In the latter case the critical questions would be employed in turn to further provide rationale for this recommendation.

Another example of a possible dialogue should Bob propose an alternative pain killer as illustrated in Fig. 11:

- *bob*: Should I take Codeine?
- *agent*: Codeine (Treatment T) should not be considered as it caused you side effects in the past.

Here the agent is using the explanation template from Fig. 6 that matches CQ2.

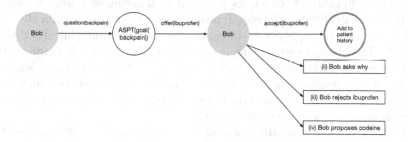

Fig. 9. Start of the dialogue protocol using approach in [30]

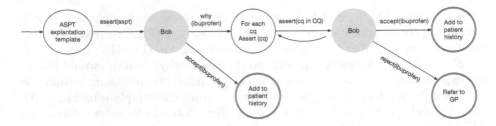

Fig. 10. (ii) Explanation dialogue: why branch of dialogue protocol and (iii) persuasion dialogue branch of dialogue protocol using approach in [30]

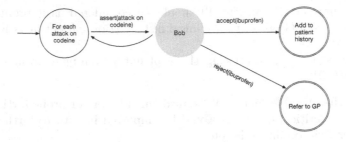

Fig. 11. (iv) Persuasion dialogue on alternative action branch of full dialogue protocol using approach in [30]

5   Related Work

This section succinctly outlines some of the work that relates to the elements proposed in this paper. This includes work on argumentation based clinical decision support, argumentation-based dialogues and explainable AI.

While "explainable AI" is a recently coined term, work on explanation has deep roots within the field of AI. As Van Lent *et. al* argue, such work goes back at least to the 1970s through the requirement of expert systems, specifically clinical ones, to be able to supply a justification for the recommendations that they made [29]. Despite this long history, it is only recently that the AI community has identified explainability as a key challenge problem [10]. For example, suggesting that it will be a requirement for the adoption of AI-based systems (such as deep learning systems) that they are able to explain the reasoning process rather than just recommending some actions to take.

The topic of what constitutes an explanation is broadly addressed by Miller *et al.* [18,19] who describes several approaches taken in the literature such as *explanation selection* where a set of reasons are chosen to explain a decision. This overview suggests several areas for investigation including *explanation as conversation*. Such an approach is closely related to what we discuss here where we model human-agent interactions for conveying explanations to users.

Turning to the more specific topic of argumentation-based (specifically argumentation scheme-based) systems for clinical decision support, we find several systems that are close to what we describe. In StAR [8,12] argumentation was used as a means to support reasoning about risk in the absence of numerical estimates of uncertainty. This was in the context of reasoning about the carcinogenic risk of different chemicals. The RAGs approach [6] leverages arguments as a way of visually representing the different reasons underpinning different possible treatment actions for a given patient. The DRAMA agent proposed in [2] uses argumentation schemes to generate arguments used to reason about a treatment, but this work does not touch on explanations or offer a dialogue based interface. Similarly the *Carrel+* project [31], where the objective of the argumentation based tool was to supervise and validate the deliberation process on organ transplant viability. In *Carrel+*, arguments were generated based on argumentation schemes and the goal was not explaining decisions but agreeing on a decision. The Argumentation based decision support approach proposed in the *CISpaces* project in [4] presents an explanation template, and there is also some groundwork for explainability from argumentation frameworks articulated in [25]. Finally, we note that [3] also points out that there is a close relationship between dialogue and explanation.

6   Conclusions

We have presented a novel approach that makes use of argumentation-based dialogues and explanation templates to implement the explanation functionality in the context of wellness consultations between an agent acting as a GP and the patient as the human. Our contributions include modeling the types of human-agent dialogues of relevance to wellness consultations, articulating an approach to providing explanations as part of an argumentation-based dialogue and illustrating the applicability of this approach on a clinical example related to patient self management.

In future work, we plan to extend this approach to additional types of interactions as sketched in Fig. 1. Here we modelled interactions where a human plays the role of patient and a software agent plays the role of GP. Other interactions include those in which both agent and human play the role of patient, and so the two interact as peers. In this work we made some assumptions on the differing level of knowledge of the user as compared to the agent, and in peer to peer interactions this would not be the case in general. Thus this future work will explore the effect of the hierarchy of levels of knowledge on the dialogue models. In addition, our plans for future work recognise that reasoning about what possible treatments should be adopted in a clinical context involves more than one argumentation scheme. Thus we will study additional schemes. Indeed, we are currently in the process of mapping the general process of deciding on treatments for stroke to existing argumentation schemes, specializing these schemes where appropriate and highlighting gaps where existing argumentation schemes do not cover the necessary reasoning. This will provide an exhaustive set of schemes covering medical treatment relevant to stroke. This exhaustive set of stroke treatment schemes will then drive the development of an associated set of explanation schemes and their associated critical questions.

These dialogues are being implemented in a chatbot, and we will evaluate the approach as part of a user study with stroke patients. This study will allow us to assess the feasibility of the approach and establish, qualitatively, how patients interact with such a system. We will be evaluating this consult app as part of a user study in the near future. The dialogue system is being developed in conjunction with a GP, but it would be desirable to test this interaction with a wider group GPs to assess whether these protocols align with the conversations GPs have with patients.

Other future work will concentrate on critical questions. In this paper we have articulated the role of critical questions in the reasoning about recommended actions and the rationale for those actions. In future work we will consider the role of critical questions both when they attack an argument and when they are satisfied and therefore do not generate an attack. Explanations provided within the dialogues should be able to explain both outcomes for each critical question. This will facilitate situations where argumentation schemes are associated with different critical questions and the answer to the critical questions may change over time.

Finally, we will consider alternative ways to create arguments. Here we assumed that all the arguments supporting a recommended action or treatment were constructed using the clinical-specific argumentation scheme ASPT, and we have already discussed expanding the set of argument schemes to cover a wider set of situations. However, this is not the only way to create arguments. For example, arguments can be mined from data [25,28], and, in the long term, we would like to see the arguments used by a system like ours being mined from the medical literature.

Acknowledgements. This work was funded by EPSRC grant EP/P010105/1 CON-SULT: Collaborative Mobile Decision Support for Managing Multiple Morbidities.

References

1. Atkinson, K., Bench-Capon, T.: Practical reasoning as presumptive argumentation using action based alternating transition systems. Artif. Intell. **171**(10–15), 855–874 (2007)
2. Atkinson, K., Bench-Capon, T.J.M., Modgil, S.: Argumentation for decision support. In: Database and Expert Systems Applications (DEXA), pp. 822–831 (2006)
3. Bex, F., Walton, D.: Combining explanation and argumentation in dialogue. Argument Comput. **7**(1), 55–68 (2016)
4. Cerutti, F., Norman, T.J., Toniolo, A., Middleton, S.E.: CISpaces.org: from fact extraction to report generation. In: Proceedings of COMMA 2018 Computational Models of Argument, pp. 269–280 (2018)
5. Cogan, E., Parsons, S., McBurney, P.: What kind of argument are we going to have today? In: Proceedings of the 4th International Joint Conference on Autonomous Agents and Multiagent Systems, pp. 544–551. ACM (2005)
6. Coulson, A., Glasspool, D., Fox, J., Emery, J.: RAGs: a novel approach to computerized genetic risk assessment and decision support from pedigrees. Methods Inf. Med. **40**(4), 315–322 (2001)

7. Dung, P.M.: On the acceptability of arguments and its fundamental role in non-monotonic reasoning, logic programming and n-person games. Artif. Intell. **77**(2), 321–357 (1995)
8. Fox, J.: Will it happen? can it happen? a new approach to formal risk analysis. Risk Decis. Policy **4**(2), 117–128 (1999)
9. Girle, R.A.: Commands in dialogue logic. In: Gabbay, D.M., Ohlbach, H.J. (eds.) FAPR 1996. LNCS, vol. 1085, pp. 246–260. Springer, Heidelberg (1996). https://doi.org/10.1007/3-540-61313-7_77
10. Gunning, D.: Explainable Artificial Intelligence (XAI). Defense Advanced Research Projects Agency (DARPA) (2017)
11. Kokciyan, N., et al.: Towards an argumentation system for supporting patients in self-managing their chronic conditions. In: Proceedings of the AAAI Joint Workshop on Health Intelligence (2018)
12. Krause, P., Fox, J., Judson, P., Patel, M.: Qualitative risk assessment fulfils a need. In: Hunter, A., Parsons, S. (eds.) Applications of Uncertainty Formalisms. LNCS (LNAI), vol. 1455, pp. 138–156. Springer, Heidelberg (1998). https://doi.org/10.1007/3-540-49426-X_7
13. McBurney, P., Parsons, S.: Chance discovery using dialectical argumentation. In: Proceedings of the Workshop on Chance Discovery, Fifteenth Annual Conference of the Japanese Society for Artificial Intelligence. Matsue, Japan (2001)
14. McBurney, P., Hitchcock, D., Parsons, S.: The eightfold way of deliberation dialogue. Int. J. Intell. Syst. **22**(1), 95–132 (2007)
15. McBurney, P., Parsons, S.: Representing epistemic uncertainty by means of dialectical argumentation. Ann. Math. Artif. Intell. **32**(1–4), 125–169 (2001)
16. McBurney, P., Parsons, S.: Games that agents play: a formal framework for dialogues between autonomous agents. J. Logic Lang. Inf. **11**(3), 315–334 (2002)
17. McBurney, P., Parsons, S.: A denotational semantics for deliberation dialogues. In: Proceedings of the 3rd International Conference on Autonomous Agents and Multi-Agent Systems. IEEE Press (2004)
18. Miller, T.: Explanation in artificial intelligence: Insights from the social sciences. Artif. Intell. **267**, 1–38 (2018). ISSN 0004-3702. https://doi.org/10.1016/j.artint.2018.07.007. http://www.sciencedirect.com/science/article/pii/S0004370218305988
19. Miller, T., Howe, P., Sonenberg, L.: Explainable AI: beware of inmates running the asylum. In: Proceedings of the IJACI Workshop on Explainable AI (2017)
20. Parsons, S., et al.: Argument schemes for reasoning about trust. Argument Comput Spec. Issue Trust Argum. Technol. **5**(2–3), 160–190 (2014)
21. Parsons, S., McBurney, P., Sklar, E., Wooldridge, M.: On the relevance of utterances in formal inter-agent dialogues. In: Proceedings of the 6th International Conference on Autonomous Agents and Multiagent Systems (2007)
22. Parsons, S., Wooldridge, M., Amgoud, L.: Properties and complexity of some formal inter-agent dialogues. J. Log. Comput. **13**(3), 347–376 (2003)
23. Parsons, S., Wooldridge, M., Amgoud, L.: On the outcomes of formal inter-agent dialogues. In: Proceedings of the 2nd International Conference on Autonomous Agents and Multi-Agent Systems. ACM Press, New York (2003)
24. Prakken, H.: Formal systems for persuasion dialogue. Knowl. Eng. Rev. **21**(02), 163–188 (2006)
25. Rago, A., Cocarascu, O., Toni, F.: Argumentation-based recommendations: Fantastic explanations and how to find them. In: Proceedings of the 27th International Joint Conference on Artificial Intelligence, pp. 1949–1955 (2018)

26. Rahwan, I., Ramchurn, S.D., Jennings, N.R., Mcburney, P., Parsons, S., Sonenberg, L.: Argumentation-based negotiation. Knowl. Eng. Rev. **18**(4), 343–375 (2003)
27. Rahwan, I., Simari, G.R.: Argumentation in Artificial Intelligence, vol. 47. Springer, Heidelberg (2009)
28. Rajendran, P.: Aggregating and Analysing Opinions for Argument-based Relations. Ph.D. thesis, University of Liverpool, Liverpool, June 2019
29. Shortliffe, E.H., Davis, R., Axline, S.G., Buchanan, B.G., Green, C.C., Cohen, S.N.: Computer-based consultations in clinical therapeutics: explanation and rule acquisition capabilities of the MYCIN system. Comput. Biomed. Res. **8**(4), 303–320 (1975)
30. Sklar, E.I., Azhar, M.Q.: Argumentation-based dialogue games for shared control in human-robot systems. J. Hum. Robot Interact. **4**(3), 120–148 (2015)
31. Tolchinsky, P., Cortes, U., Modgil, S., Caballero, F., Lopez-Navidad, A.: Increasing human-organ transplant availability: argumentation-based agent deliberation. IEEE Intell. Syst. **21**(6), 30–37 (2006)
32. Walton, D., Krabbe, E.C.W.: Commitment in Dialogue: Basic Concepts of Interpersonal Reasoning. State University of New York Press, Albany (1995)
33. Walton, D., Reed, C., Macagno, F.: Argumentation Schemes. Cambridge University Press, Cambridge (2008)

Explainable AI and Cognitive Science

A Historical Perspective on Cognitive Science and Its Influence on XAI Research

Marcus Westberg[1(✉)], Amber Zelvelder[2], and Amro Najjar[2]

[1] Department of History, Philosophy and Religion,
Oxford Brookes University, Oxford, UK
`westberg.m@gmail.com`
[2] Computer Science Department, Umea University, Umea, Sweden
`{amberz,najjar}@cs.umu.se`

Abstract. Cognitive science and artificial intelligence are interconnected in that developments in one field can affect the framework of reference for research in the other. Changes in our understanding of how the human mind works inadvertently changes how we go about creating artificial minds. Similarly, successes and failures in AI can inspire new directions to be taken in cognitive science. This article explores the history of the mind in cognitive science in the last 50 years, and draw comparisons as to how this has affected AI research, and how AI research in turn has affected shifts in cognitive science. In particular, we look at explainable AI (XAI) and suggest that folk psychology is of particular interest for that area of research. In cognitive science, folk psychology is divided between two theories: theory-theory and simulation theory. We argue that it is important for XAI to recognise and understand this debate, and that reducing reliance on theory-theory by incorporating more simulationist frameworks into XAI could help further the field. We propose that such incorporation would involve robots employing more embodied cognitive processes when communicating with humans, highlighting the importance of bodily action in communication and mindreading.

Keywords: XAI · Cognitive science · Folk psychology

1 Introduction

Philosophy has had many influences on cognitive science, especially in regards to views on the nature of the mind. How we understand the mind affects how we seek to construct artificial intelligence. Jerry Fodor's computational theory of mind [21] has served as a platform for AI research as it states that the workings of the human mind are fundamentally algorithmic manipulation of symbols and thus perfectly possible to recreate in an artificial environment. Similarly, embodied approaches to cognition have had positive effects on robotics, as theories of

© Springer Nature Switzerland AG 2019
D. Calvaresi et al. (Eds.): EXTRAAMAS 2019, LNAI 11763, pp. 205–219, 2019.
https://doi.org/10.1007/978-3-030-30391-4_12

the mind focused on world-agent interaction have inspired reactive bottom-up systems allowing robots to navigate the world through much less data-hungry decision making. In this article, the term "theory of mind" refers specifically to a held view or theory within philosophy or cognitive science that proposes an explanation to how the mind works, and not the agent-specific ability to attribute mental states to others, as the term is often used in relation to folk psychology. So, what are the fundamental assumptions of the mind that can be found in machine learning and explainable AI, and could philosophy or cognitive science help improve these assumptions? Since the field is concerned with human-robot interactions and understanding, there has to be one or several assumptions about how mindreading occurs and what role different types of interactions play in the cognitive processes leading to forming such understanding. There would also be underlying theories about how we rationalise and structure our internal models of the world (if at all) and how this comes into play when describing our intentions and actions. This in turn highlights the problem of the evidentiary boundary between mind and world, both robotic and human, and how we overcome it. Philosophy offers many different solutions to this problem, some elevate the role of action and behaviour in perception over internal representation in order to bypass the problem altogether, such as enactivism, while others remain strictly internalist but propose predictive models in order to unify world and agent through Bayesian best-performing hypotheses, something that can be found in predictive processing and predictive theories of mind. By looking at robots/AI, we do not only see assumptions about how robot minds should work, but also about how human minds interact with the world and other agents within it. In order for AI to be understandable in such a way that humans can cooperate with it, we need these theories to be compatible so that we can bypass the evidentiary boundary between what we see in the world (an agent's behaviour and outward communication) and what is going on with the agent's mental states (intentions, beliefs, possible future actions).

This article highlights the intertwined relationship between cognitive science and AI, and reviews how the history of philosophy of mind can be echoed in the history of developments in approaches to AI research (Sect. 2), then it identifies some examples of cognitive science-inspired AI application (Sect. 3), discusses folk psychology (Sect. 4) and presents a roadmap describing how XAI can profit from the recent evolution in theories of mind (Sect. 5). Finally, Sect. 6 concludes this article.

2   Tracing the History of Philosophy of Mind and Cognition

The history of AI research is often directly or indirectly linked to the history of cognitive science and theories of mind. Theories about how the human mind works can inform AI research in what to model their work on, and successes and failures within AI can teach us lessons about of the human mind itself is likely to work. If we were ever to create a general-level artificial intelligence that is perfectly indistinguishable from a human in cognitive capacities, we would likely want to say that we have also gained some insight into a plausible model of the human mind. Similarly, should theorized models of how the human mind works from a field such as philosophy fail as models for creating efficient and functional AI, then there is an equally compelling argument to be made that such results discredit said theorized models. An example of this crossover effect can be seen in the history of the Computational Theory of Mind (CTM). The Computational Theory of Mind was born out of the emergence of computing machines, most prominently the abstract Turing machine, invented by Alan Turing [59] as a model to, among other purposes, find a solution to the 'decision problem' in formal logic. A line of thought came about that if machines can perform calculations and solve problems, could they be considered intelligent? Could a machine think? Turing himself asked these questions [60] and developed what would be known as the Turing test, a test where an interrogator would pass written questions to two other anonymous participants (one human, one machine) and would receive answers from both. The interrogator would then determine which participant is the machine, and which is the human. The purpose of the test was for the machine to exhibit their capacity of intelligent behaviour. If the machine could, even after rigorous questioning, appear to the interrogator as the human participant, then the machine had a claim for being an intelligent thinker.

This pursuit of intelligent machines had a reverse side to it: If these machines could solve problems and make decisions in a way that approaches cognition, could cognition itself in fact be computational? Such a thought offered a robust and structured approach to modelling the mind, while providing an analogous relationship with the emerging science of creating better and more complex computers. The rise of a computational theory of mind took place in this landscape, beginning in the earliest stages with a suggestion by McCulloch and Pitts [51] but was properly put into theory by Putnam [56] and was further developed by Fodor [21].

2.1 Fodor and Cognitive Science

The comparison between human intelligence and the workings of computers offered a solid argument for how the mechanisms of our mind and thoughts were structured, and provided a bridge between the abstract mind models of philosophers with the physical form which these cognitive faculties inhabit. If machines could think, then human thinkers could also be like machines. CTM served as a very influential founding theory in cognitive science [52]. While the theory started with Putnam [56], it was further developed and popularized by Fodor [21]. The theory is grounded in the idea that the mind works in many ways like a digital computer; the mind is parsing internal representations (symbols) in algorithmic ways, forming an internal computational language that is used to process input data into output. Fodor saw the symbol-dependent processing of the mind as a language and referred to this internal syntax as "the language of thought" and "mentalese". This interpretation placed mentalese as essentially a computational language of the mind, physically realised in the brain. By comparison, we could compare this to how a computer language works, and how the symbols and syntax of programming languages are realised in software (thinking), but physically situated in hardware (the brain). In fact, Fodor's theory claimed that this is exactly what is going on, suggesting that the mind is a physically realized computational environment where information processing occurs. That is, our minds are a formalized system parsing a language based on information-carrying representations, structured by syntactic and semantic rules, the upshot of this idea being that the mind and the world are connected through our understanding of what is effectively a second simulated world within our mind. Information about the world enters the mind as sensory data; the things we see, hear, taste, smell and feel all enter our mind as raw data, which is then used to form mental representations, starting simple and building in complexity to form concepts. These representations are the components of our thought processes, which in turn are algorithmic in nature; our thought processes are problem-solving operations using an internal rule set which determines how the symbols (representations) are to be manipulated by the system. This representation in turn can vary in complexity and structure, such that they may be structurally atomic or molecular, which then carries down to their syntactic constituents [23].

CTM became a very popular theory of mind and gave new fuel to cybernetics (which had already been around since the 1930s) which led to the formation of modern cognitive science, as well as the resurgence of artificial intelligence research. As a result of this, throughout the 70s, 80s and 90s, much of the research in cognitive science and AI followed these computational mind models. In neuroscience as well, CTM was adopted by David Marr as a computational theory of vision [43,44]. However, the 90s were a decline in CTM's era as the leading theory of mind. One of the reasons was the rise in popularity of connectionism, and the people adopting it generally went against the idea of a language of thought. Connectionist models of the mind are built upon neural networks of interconnected nodes rather than the more linguistically inspired CTM. In particular, eliminative connectionism sought to move away from the idea of computationalism and mental representation in thought [14,37]. Thus the 90s was a transformative era where philosophers started to move away from the idea of classical CTM and arrived in a new paradigm era where the classical computational theory of mind as offered by Putnam and Fodor had decreased in popularity.

2.2   Embodied Cognition and Robotics

While the linguistic aspects of the classical version of CTM declined in the face of connectionism and research into neural network architectures, CTM's input-output model of mind also faced criticism for its inefficiency in how it modeled our interaction with the world. Classical CTM relied heavily on large amounts of data being collected in order for the mind to structure and learn about the world. This was also true of Marr's theory of vision although there were discussions in place that addressed the issue of dealing with uncertainty in identifying partial objects [44]. CTM had traditionally constructed our interaction of the world as being input-output, that is we perceive the world as it enters our senses (input), the sensory data transforming into a mental image of the world within our mind, which we then act upon (output). This has sometimes been referred to as the input-output sandwich, the idea being that cognition is filling, trapped between the bread of perception and action. Such a model, in order to connect with the outside world, first has to gather enough data via perception in order to create a working understanding of the world before being able to make decisions about actions to be taken within that world. Such systems are information-hungry in ways that human-like animals often could not afford to be, which was considered a flaw in classical CTM's plausibility as a model for how our minds actually work. Similarly, AI that relied on intensive representation processing was still far away from achieving something akin to what a human mind could do, and often required quite complex top-down systems to guide their decision-making. Meanwhile, robotics was starting to see new successes by using motion and the world integrate with their cognitive systems [45,46] and roboticist Rodney Brooks presented a new type of computational architecture that was light on representation crunching and more focused on world-driven processes [8,9]. Philosopher Andy Clark [15] identified this new movement as the rise of Embodied Cognition, proposing a model of the mind where bodily action was more integrated with the classically introverted and isolated cognition presented in CTM. This theory of mind allowed for cognition to offload processes onto the world, and to use the environment for cognitive scaffolding. This paradigm shift made cognitive research more focused on mind-world-body interplay and interaction rather than complex internal architectures. For example, Collins et al. performed research into passive dynamic walker robots that use the natural pendulum motions of legs to aid walking, creating a smoother gait than the computation heavy and precision-demanding alternatives [18]. In philosophy, these developments led to both CTM and connectionism losing popularity and attention in favor of the now dominant EEEE theories: Embodied, Embedded, Extended and Enactive cognition.

2.3   Predictive Processing Against Radical Enactivism

As classical CTM loses influence, philosophers in the early 2000s and onwards become increasingly interested in describing the mind through world-driven processes and much interest is given to where we draw the border between mind and world. Classical CTM was traditionally very brain-focused in this regard, but the new EEEE theories all have in common that they attempt to expand this view, or reject it entirely. Embodied cognition [15] as mentioned before pays attention to how the body can play a causal role in cognition; embedded cognition involves the usage of external tools to facilitate cognitive processes; extended cognition is a functionalist stance that argues that external processes, if they fulfil the same functional role as the processes our heads, would constitute as part of our mind [17]; enactivism proposes the idea that experience of the world is conceived by interaction between brain, body and world, thus making cognition a dynamic activity rather than a passive intake of information [62].

The elements of these theories have branched off in many ways, and although the general emphasis within philosophy these days lies in 'active' cognition, i.e. focused on world-engaging processes rather than the more 'passive' traditionalist model, many positions have appeared that challenge each other.

The discovery of backward connections in the brain and an increased interest in Bayesian prediction models has given rise to theories of mind based on prediction error minimization in perception called predictive processing, where the mind meets the world by predicting future input. Some use this theory to defend the traditionalist view of a brain-centric mind [36] while others seek to bridge the gap between embodied, enactive and representational models [16]. Either of these approaches to predictive processing still support the idea of mental representation, something that was core to classical CTM, the effective change being that predictive processing has taken care of the problem of inefficient input ecology. On the other hand, branches of enactivism such as radical enactivism argues against the existence of mental representations or content [38], claiming that enactive processes make contentful representations functionally obsolete. This creates a new conflict within philosophy of mind where our fundamental understanding of the mind has changed from a passive observer to an active one, while the debate about representations in cognition continues much like it did during the rise of eliminative connectionism.

2.4   Influences in XAI

As noted above, modeling an artificial intelligence can be influenced by how we as a scholastic community understand minds to work, be that as classical CTM, as predictive or embodied. However, eXplainable AI (XAI) is not only concerned with how minds work but also with mindreading and a cognitive agent's ability to explain itself to other minds. How this explanation is structured is related to what theory of mind we endorse. An AI modeled after classical CTM would be primarily concerned with reading and communicating its cognition in terms of representational mental states, while an AI incorporating embodied cognitive processes could make use of more world-driven processes in order to both read its audience and explain itself. For example, this could involve not only analyzing motion, facial expressions and other external signs of mental states, but in turn using these to both in performing cognitive processes and communicating said cognitive processes. The kind of interaction proposed here specifically relates to incarnate agents where human and AI communication occurs through means of folk psychology and expression of intention. As such, this is not as relevant to systems that utilise other forms of communication such as diagnostics. Additionally, XAI is in the position where we are not only invested in how to structure the mind, but how we as agents understand other minds to work. In this way, XAI is influenced by and concerned with folk psychology, which has an extensive history within philosophy of mind and cognitive science. Concerns to be raised about folk psychology are thus also concerns relevant to XAI. In this paper, the term folk psychology is used specifically to refer to the cognitive capacity to (a) attribute and explain mental states in other cognitive agents, (b) predict future behaviour of other cognitive agents and (c) to manipulate or coordinate behaviour with other cognitive agents through this attribution and prediction of mental states [35]. Broadly, possessing these capacities means that we describe the actions of others in intentional terms, highlighting the fact that we view other agents not as mere objects of causality, but as minds with their own beliefs and desires [57]. While these capacities are traditionally talked about in the context of relating to other human beings, there is a broader sense in which such capacities could be directed toward reading other functionally similar minds, for example reading and attributing mental states to non-human animals, or an AI.

3 Cognitive Science Inspiring AI Applications

In order to highlight the impact of developments in cognitive science of AI, this section presents examples of recent AI applications relying on enactivism (Sect. 3.1), and discusses how the renewed interest in XAI has relied on theories of mind (Sect. 3.2).

3.1 Enactive and Developmental AI

Enactive AI is inspired from the works on enaction developed by the cognitive biologists Maturana and Varela [47,48]. In contrast to the cognitivism of the traditonal approaches, which involves a view of cognition that requires the representation of a given objective pre-determined world [61,63], enaction relies on the assumption that existence of a cognitive agent are *enacted* (*i.e.* co-determined) by the agent as it interacts with its environment within which it is embedded. Thus, nothing is predetermined, and cognition becomes the process whereby an autonomous system becomes viable and effective in its environment [63]. Co-determination means that, on the one hand, the agent is specified by the environment and, on the other hand, it is the cognitive process itself which determines what in the environment is real and meaningful for the agent.

Since Enaction considers that the construction of cognition is undertaken on the basis of interactions between the agent and their physical and social environments, it supports constructivism, self-organization and developmental agents [19].

In contrast to the classical approach and the computational theory of mind, AI architectures based on enaction allow to overcome well known problems such as the *frame problem* [50], the *symbol grounding* problem [33], and *modeling of common-sense* problem [50].

Moreover, Enaction based AI is appealing because it allows Enactive-AI system to retain the following three characteristics [19]: *(i) no need for a-priori representations*: the agent does not need to have a pre-given model of its world. Instead, it can learn its environment when it interacts with it. *(ii) plasticity*: the agent is capable of adapting to the its environment even when significant distributions take place. This plasticity is located both at the physical (bodily) interactions (e.g. a robot capable of adapting its movement to an unforeseen slippery ground), and at the nerve level of higher interactions (*cerebral plasticity*). *(iii) co-evolution*: A modification of the world by the agent in return imposes a modification of that agent.

The characteristics mentioned above made enactive and developmental approaches to AI appealing for many AI applications, including developmental robotics [4], smart environment [49,53], and road-traffic control [31].

3.2   The Case for Explainable AI

Since the turn of the century, intelligent applications are becoming more and more pervasive in our daily lives. As these applications get more and more sophisticated, there is an impelling need to make them explainable. This tendency is accentuated by the rise of black-box machine learning mechanisms [32] (e.g. deep learning) and their, sometimes, intriguing results [58]. To overcome these problems, recent legislation in the EU, emphasized the right of explanation [64]. Furthermore, evidence from user studies suggest that humans tend to anthropomorphize intelligent systems and attribute them with a State of Mind (SoM) [35]. This tendency is known as the *intentional stance* [20], and it pushes humans to explain the behavior of these systems in terms of beliefs, goals, and sometimes emotions [35]. For these reasons, recent research on XAI gained a significant new momentum [2]. XAI aims to offer explanations that would help the user to understand intelligent system and would lead to better human-agent collaboration and incite the user to understand the capabilities and the limits of the system, thereby improving the levels of trust and safety, and avoiding failures, since the lack of appropriate mental models and knowledge about the system may lead to failed interactions [3,11]. XAI is still in its early stages of development, for this reasons, most existing works are either carried out at the conceptual front [2]. Systems based on the Belief-Desire-Intention (BDI) architecture constitute a significant portions the few applied works [5]. The fact that the BDI agent architecture is inspired from Folk Psychology [55] makes it suitable to explain the agent intentions to the lay user [7]. Next section presents folk psychology and discusses its relationship with XAI.

4   Folk Psychology and XAI

In philosophy and cognitive science there are two main theories for how mindreading (i.e. folk psychology) takes place: theory-theory and simulation theory. In Sect. 4.1 we will introduce theory-theory, followed by simulation theory in Sect. 4.2. The basic distinction to be made is that theory-theory holds the view that folk psychology operates from a set of rules that we hold in our mind, a distinct theory about mental states in others that informs our interpretation. Simulation theory on the other hand claims that our mindreading is performed by mental simulation, changing our perspective to that of others so that we may come to understand the underlying mental states for their behaviour.

4.1   Theory-Theory and Classical CTM

Theory-theory proposes that our understanding of other minds is built upon a tacit theory that we possess about how fears, desires and other mental states operate in other human beings, including the causal relationship these mental states have in a social context (i.e. how anger in one person can cause another to become sad, what it means to be sad and what kind of behaviours signify a mental state of sadness, how this sadness can be alleviated etc.). When we interact with or observe other cognitive agents, we employ this theory as a framework through which we form an understanding of the mental states in these agents based on what we can perceive in their behaviour. For example, person A sees person B reach out their hand toward an apple hanging on a branch. Person A employs their theory that people who reach out their hands do so out of a desire to grab what their hand is reaching toward. Since an apple is an edible object, person A further projects that person B is likely to be hungry, since hunger drives desire for edible things. As such, person A has now mindread a state of hunger in person B, as well as their desire for an apple. Person A can then also predict future behaviour in person B, in that once they get the apple are likely to proceed with eating it in the near future.

There is significant overlap and compatibility between the theory-theory and classical CTM, especially when considering Fodor's theories on modularity of mind, where similarly tacit cognitive subsystems, i.e. modules, operate on a contextual basis for specific purposes - for example vision and language acquisition [22]. In a similar vein, theory-theory can be described as a folk psychology module in the human mind - a special capacity that is normally not part of our central processing, but within the right context receives the input (the perception or information of actions and behaviour of external agents), interprets the information through our folk psychology ruleset and produces an interpretation of what mental states we are perceiving. Older versions of theory-theory also tend toward a sentence-based representational structure closely resembling that of classical CTM, which is structured as a language of thought [21], but there are also those criticising this idea and proposing connectionist structure, while still maintaining the essence of a theory-based folk psychology [13,57].

Since theory-theory posits a set of rules or hypotheses informing the interpretations produced by our mindreading, this ruleset has to at some point be developed in the agent. This is an intersection where some supporters of theory-theory embrace nativism while others argue for empiricism. The nativist stance proposes that the framework for folk psychology is innately present at birth, and that its formation is part of a development pattern present in our genes [10]. Empiricists in contrast argue that the development of folk psychology is a process based on evidence-based theory formation throughout childhood and onward [28,29]. This debate can be compared to the debate of nativism versus empiricism in language acquisition, with similar arguments to be made, such as nativism's critique of the poverty of stimulus [12].

4.2   Simulation Theory

Simulation theory contrasts itself with theory-theory in claiming that our process of mindreading is largely based on our ability to put ourselves in the shoes of others, in essence we simulate the behaviour of others in our own mind as if from our own perspective. There are two components to this that need to be unpacked: (1) how mental states in other persons are simulated in our own mind, (2) how these simulated mental states serve mindreading and prediction.

When it comes to (1), some supporters of simulation theory have latched onto the discovery of mirror neurons as signifying a mirroring capacity that could be linked to mental state simulation [24]. However, the modern version of simulation theory came about before this discovery, in the mid-1980s [30]. Thus for the first decade of its popularity, and still ongoing for those who do not subscribe to the idea of mirroring, this capacity is instead described in terms of empathy or imagination.

Regarding (2), simulating mental states is not enough to predict behaviour, as there needs to be a process in place to explain how these mental states relate to possible future behaviour. What follows is thus a simulated decision-making process, where the agent asks themselves what they would do given the simulated premises. Essentially, the agent asks themselves *"what would I do if X?"* where X is the simulated mental states and context of the subject of the mindreading. The answer to that question becomes the theory used to predict the subject's future actions.

Simulation theory is thus frugal in the sense that it does not require an information-rich theory of how other minds work. Instead, all that is required is the capacity for the agent to place their frame of mind in someone else's situation through empathic simulation, and what follows is no different from the kinds of decision-making processes that take place in the agent's own actions. This simulated process greatly reduces the amount of cognitive capital spent on mindreading in comparison to theory-theory, since mindreading becomes primarily process-driven rather than theory-driven [25].

5   A Roadmap for XAI

As mentioned before in Sect. 2.4, folk psychology is relevant to XAI research. As XAI involves both humans explaining themselves to AI and, more importantly, AI explaining itself to humans, mindreading becomes an important aspect of this exchange of information and understanding. Understanding robots as subjects of mindreading helps creating a framework where the mental states of an AI can be explained, and thus aids the strategies employed by the AI in explaining itself to humans. These influences of folk psychology can already be found in XAI. For example, belief-desire-intention (BDI)-based agents [7,39] are an application of theory-theory in XAI: These agents have their actions explained through beliefs and goals, generating a theory of future behaviour based on an understanding of these beliefs and goals, and how they relate to actions taken. BDI agents are widely used for social simulations [1]. BDI agents offer a reasoning formalization inspired by human mentality based on intuitive concepts that allow for a straightforward implementation in IT systems. Hence, the BDI architecture has been highlighted as a practical solution to model humans and create human-like behavior in simulated environments [55].

However, the way these are communicated to humans, and thus the method through which a human agent is invited to mindread the BDI-based agent, is through either natural language explanation, or an understanding of the BDI-based agent's goal hierarchy tree (GHT). This kind of explanation involves no amount of simulation or putting oneself in the robot's shoes, but rather becomes a task of piecing together the robot's reasoning through empirical questioning and investigation, thus making it a clear case of theory-theory.

While this may be functionally sufficient in allowing the robot to explain itself within the context of its design, it is limiting in the sense that underutilises the human capacity for empathetic mentalisation, specifically when it comes to mindreading through interpreting action - something that both human children and chimpanzees have been shown capable of [34]. This appeal to empathetic mentalisation is a trait of simulation theory. Applying simulation theory as a framework for XAI could thus upon up a broader scope for communication by capturing more of the human experience.

Simulation theory, as presented in Sect. 4.2, possesses a greater advantage over theory-theory in that it can bypass the nativist versus empiricist debate present in theory-theory. The ability to learn mindreading from simulation provides an innate system for learning where our knowledge and understanding of other minds does not have to come pre-constructed at birth, nor does the learning itself involve the construction of a rigorous ruleset. Instead, any constructed rulesets would be created through best-making hypotheses much in the manner of a prediction-error minimization model for perception [36], where the more advanced facets of our knowledge of human behaviour are constructed models from a much more basic and frugal process.

However, it is important to note that theory-theory is improving by making adjustments to incorporate Bayesian prediction models, just like predictive processing revitalised the representational aspects of CTM in the face of increasingly non-representational alternatives [16]. In theory-theory's case, this turn toward prediction is pertaining only to the empiricist part of the divide, and thus not a nativist strategy [27]. This, however, may not be enough for standalone theory-theory to win out over simulation theory.

Instead, what is becoming increasingly popular are hybrid theories, incorporating simulationist elements, but falling back on theory-theory wherever simulation alone is insufficient [26] or vice versa [54]. Such hybrid theories would still go well together with predictive processing: Theories that are primarily simulationist with elements of theory-theory would benefit from prediction error minimization's explanation of how theory can be constructed from continuous and updating simulation, while primarily theory-based theories with elements of simulation will find an increase in cognitive frugality that explains away the poverty of stimulus argument while avoiding nativism, thus promoting bottom-up over top-down learning. Thus even if standalone simulation theory does not stand as a clear superior theory, there is still a strong argument to be made that including elements of simulation theory presents a stronger theory for how we mindread over a standalone theory-theory. If this holds true, then it is even more important that XAI going onward incorporates more elements of understanding the AI as an embodied agent.

Some XAI research already shows an adaptation and understanding of the importance of a robot's physical movements and how nuances in this affect how a human observer interprets its mental states [35]. Experiments involving robots communicating their intentions with bodily movements [42] or their mental state through eye movements and posture [6] show positive results in human interaction with robots and opens up a new path of interpretation. Furthermore, the importance of emotions in explanations has been studied by recent works in XAI. Yet, most of these works (e.g. [40,41]) rely on BDI agents (theory-theory) and address simplified scenarios. This research shows an opening for simulation theory in XAI since simulation theory helps promote emphatic behavior (by putting ourselves in the others' shoes), and allows the agents to explain its behavior using body cues and non-verbal communications.

6   Conclusion

In this article we have shown how intertwined the relationship between cognitive science and AI is, and have reviewed how the history of philosophy of mind can be echoed in the history of developments in approaches to AI research. For these reasons, it is important for XAI to pay attention to developments in cognitive science, particularly regarding folk psychology, as these developments could inform better approaches for XAI in the future. Conversely, the successes and failures of XAI could in turn influence how the strategies employed are viewed in discussions on folk psychology.

There is a clear shift in cognitive science and philosophy of mind toward world-driven and embodied explanation of cognition, be that in regards to perception, folk psychology, decision-making or action, and this shift does not seem to be going back. Thus, the question is no longer about the input-output sandwich and how isolated thinking is from the world. Rather, the new focus is on the relationship between thinking and world. With this in mind, our prediction is that human-robot interaction will increasingly involve embodied cognition as a tool for communication. Embodied processes in robots not only act as cognitive scaffolding in navigating and interacting with the world, but it also opens them up for mindreading as they present their mental states through behaviour that humans recognise and can understand, thus allowing humans to tacitly recognise robots as cognitive agents with beliefs and desires through the same methods as they would other humans and non-human animals.

References

1. Adam, C., Gaudou, B.: Bdi agents in social simulations: a survey. Knowl. Eng. Rev. **31**(3), 207–238 (2016)
2. Anjomshoae, S., Najjar, A., Calvaresi, D., Framling, K.: Explainable agents and robots: results from a systematic literature review. In: 18th International Conference on Autonomous Agents and Multiagent Systems (2019, to appear)
3. Bethel, C.L.: Robots Without Faces: Non-verbal Social Human-robot Interaction. Ph.D. thesis, Tampa, FL, USA (2009). aAI3420462
4. Blank, D., Kumar, D., Meeden, L., Marshall, J.B.: Bringing up robot: fundamental mechanisms for creating a self-motivated, self-organizing architecture. Cybern. Syst. Int. J. **36**(2), 125–150 (2005)
5. Bratman, M.: Intention, Plans, and Practical Reason, vol. 10. Harvard University Press Cambridge, Cambridge (1987)
6. Breazeal, C., Fitzpatrick, P.: That certain look: Social amplification of animate vision. AAAI Technical Report, November 2001
7. Broekens, J., Harbers, M., Hindriks, K., van den Bosch, K., Jonker, C., Meyer, J.-J.: Do you get it? user-evaluated explainable BDI agents. In: Dix, J., Witteveen, C. (eds.) MATES 2010. LNCS (LNAI), vol. 6251, pp. 28–39. Springer, Heidelberg (2010). https://doi.org/10.1007/978-3-642-16178-0_5
8. Brooks, R.A.: Intelligence without reason. In: Proceedings of the Twelveth Internationl Joint Conference on Artificial Intelligence, pp. 569–595 (1991)
9. Brooks, R.A.: Intelligence without representation. Artif. Intell. **47**, 139–159 (1991)
10. Carruthers, P.: Simulation and self-knowledge: a defence of the theory-theory. In: Carruthers, P., Smith, P.K. (eds.) Theories of Theories of Mind, pp. 22–38. Cambridge University Press, Cambridge (1996)
11. Chandrasekaran, A., Yadav, D., Chattopadhyay, P., Prabhu, V., Parikh, D.: It takes two to tango: towards theory of ai's mind. arXiv preprint arXiv:1704.00717 (2017)
12. Chomsky, N.: A Review of BF skinner's verbal behavior. Language **35**(1), 26–58 (1959)
13. Churchland, P.M.: Folk psychology and the explanation of human behavior. Philos. Perspect. **3**((n/a)), 225–241 (1989)
14. Churchland, P.M.: A Neurocomputational Perspective: The Nature of Mind and the Structure of Science. MIT Press, Cambridge (1989)
15. Clark, A.: Being There: Putting Brain, Body, and World Together Again. MIT Press, Cambridge (1997)
16. Clark, A.: Surfing Uncertainty: Prediction, Action, and the Embodied Mind. Oxford University Press, Oxford (2016)
17. Clark, A., Chalmers, D.J.: The extended mind. Analysis **58**(1), 7–19 (1998)
18. Collins, S.H., Wisse, M., Ruina, A.: A three-dimensional passive-dynamic walking robot with two legs and knees. Int. J. Robot. Res. **20**(7), 607–615 (2001)
19. De Loor, P., Manac'h, K., Tisseau, J.: Enaction-based artificial intelligence: toward co-evolution with humans in the loop. Minds Mach. **19**(3), 319–343 (2009)
20. Dennett, D.C.: The Intentional Stance. MIT press, Cambridge (1989)
21. Fodor, J.A.: The Language of Thought. Harvard University Press, Cambridge (1975)
22. Fodor, J.A.: The Modularity of Mind. MIT Press, Cambridge (1983)
23. Fodor, J.A., Pylyshyn, Z.W.: Connectionism and cognitive architecture. Cognition **28**(1–2), 3–71 (1988)

24. Gallese, V., Goldman, A.: Mirror neurons and the simulation theory of mind-reading. Trends Cogn. Sci. **2**(12), 493–501 (1998)
25. Goldman, A.: Interpretation psychologized. Mind Lang. **4**(3), 161–85 (1989)
26. Goldman, A.I.: Simulating Minds: The Philosophy, Psychology, and Neuroscience of Mindreading. Oxford University Press, Oxford (2006)
27. Gopnik, A.: The theory theory 2.0: probabilistic models and cognitive development. Child Dev. Perspect. **5**(3), 161–163 (2011)
28. Gopnik, A., Meltzoff, A., Kuhl, P.: The scientist in the crib: minds, brains and how children learn. J. Nerv. Ment. Dis. 189 (2001)
29. Gopnik, A., Wellman, H.M.: The theory theory. In: Hirschfeld, L.A., Gelman, S.A. (eds.) Mapping the Mind: Domain Specificity in Cognition and Culture, pp. 257–293. Cambridge University Press, Cambridge (1994)
30. Gordon, R.M.: Folk psychology as simulation. Mind Lang. **1**(2), 158–71 (1986)
31. Guériau, M., Armetta, F., Hassas, S., Billot, R., El Faouzi, N.E.: A constructivist approach for a self-adaptive decision-making system: application to road traffic control. In: 2016 IEEE 28th International Conference on Tools with Artificial Intelligence (ICTAI), pp. 670–677. IEEE (2016)
32. Guidotti, R., Monreale, A., Ruggieri, S., Turini, F., Giannotti, F., Pedreschi, D.: A survey of methods for explaining black box models. ACM Comput. Surv. (CSUR) **51**(5), 93 (2018)
33. Harnad, S.: The symbol grounding problem. Phys. D: Nonlinear Phenom. **42**(1–3), 335–346 (1990)
34. Harris, P.F.: From simulation to folk psychology: the case for development. Mind Lang. **7**(1–2), 120–144 (1992)
35. Hellström, T., Bensch, S.: Understandable robots : what, why, and how. Paladyn - J. Behav. Robot. **9**(1), 110–123 (2018)
36. Hohwy, J.: The Predictive Mind. Oxford University Press, Oxford (2013)
37. Horgan, T.E., Tienson, J.L.: Connectionism and the Philosophy of Psychology. MIT Press, Cambridge (1996)
38. Hutto, D.: Enactivism: Why be radical? In: Sehen und Handeln, pp. 21–44. De Gruyter Akademie Forschung, January 2011
39. Kaptein, F., Broekens, J., Hindriks, K., Neerincx, M.: Personalised self-explanation by robots: the role of goals versus beliefs in robot-action explanation for children and adults. In: 2017 26th IEEE International Symposium on Robot and Human Interactive Communication (RO-MAN), pp. 676–682, August 2017
40. Kaptein, F., Broekens, J., Hindriks, K., Neerincx, M.: The role of emotion in self-explanations by cognitive agents. In: 2017 Seventh International Conference on Affective Computing and Intelligent Interaction Workshops and Demos (ACIIW), pp. 88–93. IEEE (2017)
41. Kaptein, F., Broekens, J., Hindriks, K., Neerincx, M.: Self-explanations of a cognitive agent by citing goals and emotions. In: 2017 Seventh International Conference on Affective Computing and Intelligent Interaction Workshops and Demos (ACIIW), pp. 81–82. IEEE (2017)
42. Kobayashi, K., Yamada, S.: Motion overlap for a mobile robot to express its mind. J. Adv. Comput. Intell. **11**, 964–971 (2007)
43. Marr, D.: Visual information processing: the structure and creation of visual representations. Philos. Trans. Royal Soc. London. B Biol. Sci. **290**(1038), 199–218 (1980)
44. Marr, D.: Vision : A Computational Investigation Into the Human Representation and Processing of Visual Information. W. H Freeman, New York (1982)

45. Mataric, M.J.: Navigating with a rat brain: a neurobiologically-inspired model for robot spatial representation. In: Proceedings of the First International Conference on Simulation of Adaptive Behavior on From Animals to Animats, pp. 169–175. MIT Press, Cambridge (1990)
46. Mataric, M.J.: Integration of representation into goal-driven behavior-based robots. IEEE Trans. Robot. Autom. **8**(3), 304–312 (1992)
47. Maturana, H.R., Varela, F.J.: Autopoiesis and Cognition: The Realization of the Living. BSPS, vol. 42. Springer, Dordrecht (1980). https://doi.org/10.1007/978-94-009-8947-4
48. Maturana, H.R.: The organization of the living: a theory of the living organization. Int. J. Man-Mach. Stud. **7**(3), 313–332 (1975)
49. Mazac, S., Armetta, F., Hassas, S.: On bootstrapping sensori-motor patterns for a constructivist learning system in continuous environments. In: Artificial Life Conference Proceedings 14, pp. 160–167. MIT Press (2014)
50. McCarthy, J.: Programs with Common Sense. RLE and MIT computation center (1960)
51. Mcculloch, W.S., Pitts, W.: A logical calculus of the ideas immanent in nervous activity. J. Symbolic Logic **9**(2), 49–50 (1943)
52. Miller, G.A., Galanter, E., Pribram, K.H.: Plans and the Structure of Behavior. Holt, New York (1967)
53. Najjar, A., Reignier, P.: Constructivist ambient intelligent agent for smart environments. In: 2013 IEEE International Conference on Pervasive Computing and Communications Workshops (PERCOM Workshops), pp. 356–359. IEEE (2013)
54. Nichols, S., Stich, S.P.: Mindreading: An Integrated Account of Pretence, Self-Awareness, and Understanding Other Minds. Oxford University Press, Oxford (2003)
55. Norling, E.: Folk psychology for human modelling: Extending the BDI paradigm. In: Proceedings of the Third International Joint Conference on Autonomous Agents and Multiagent Systems, vol. 1, pp. 202–209. IEEE Computer Society (2004)
56. Putnam, H.: Brains and behavior. In: Butler, R.J. (ed.) Analytical Philosophy: Second Series. Blackwell, Hoboken (1963)
57. Stich, S.P., Nichols, S.: Folk psychology: simulation or tacit theory? Mind Lang. **7**(1–2), 35–71 (1992)
58. Szegedy, C., et al.: Intriguing properties of neural networks. arXiv preprint arXiv:1312.6199 (2013)
59. Turing, A.: On computable numbers, with an application to the Entscheidungsproblem. Proc. London Math. Soc. **2**(42), 230–265 (1936)
60. Turing, A.: Computing machinery and intelligence. Mind **59**, 433–460 (1950)
61. Van Gelder, T., Port, R.F.: It's about time: an overview of the dynamical approach to cognition. Mind Motion: Explor. Dyn. Cogn. **1**, 43 (1995)
62. Varela, F., Thompson, E., Rosch, E.: The Embodied Mind: Cognitive Science and Human Experience. MIT Press, Cambridge (1991)
63. Vernon, D., Furlong, D.: Philosophical foundations of AI. In: Lungarella, M., Iida, F., Bongard, J., Pfeifer, R. (eds.) 50 Years of Artificial Intelligence. LNCS (LNAI), vol. 4850, pp. 53–62. Springer, Heidelberg (2007). https://doi.org/10.1007/978-3-540-77296-5_6
64. Voigt, P., Von dem Bussche, A.: The EU General Data Protection Regulation (GDPR). A Practical Guide, 1st edn. Springer, Cham (2017). https://doi.org/10.1007/978-3-540-77296-5_6

Author Index

Abchiche-Mimouni, Nadia   3
Ahlbrecht, Tobias   129
Anjomshoae, Sule   95

Bourgais, Mathieu   147

Calvaresi, Davide   41
Correia, Filipa   22

Främling, Kary   95, 110

Galland, Stéphane   41
Gomes, Samuel   22

Hochgeschwender, Nico   77
Höhn, Sviatlana   77

Jentzsch, Sophie F.   77

Kampik, Timotheus   59
Kökciyan, Nadin   186

Lindgren, Helena   59

Madhikermi, Manik   110
Malhi, Avleen Kaur   110
Mascarenhas, Samuel   22
Melo, Francisco S.   22

Micalizio, Roberto   167
Mualla, Yazan   41

Najjar, Amro   41, 95, 205
Nieves, Juan Carlos   59

Paiva, Ana   22
Parsons, Simon   186

Sassoon, Isabel   186
Schumacher, Michael   41
Sendi, Naziha   3
Sklar, Elizabeth   186
Sormano, Samuele   167

Taillandier, Patrick   147
Torta, Gianluca   167
Tulli, Silvia   22

Vercouter, Laurent   147

Westberg, Marcus   205
Winikoff, Michael   129

Zehraoui, Farida   3
Zelvelder, Amber   205

Printed in the United States
By Bookmasters